THE BLOCKCHAIN METAVERSE

A BEGINNER'S GUIDE TO VIRTUAL REALITY, AUGMENTED REALITY, DIGITAL CRYPTOCURRENCY, NFTS, GAMING, VIRTUAL REAL ESTATE, AND INVESTING IN THE METAVERSE

DONNY WALTERS

"A speech with magical force. Nowadays, people don't believe in these kinds of things. Except in the Metaverse, that is, where magic is possible. The Metaverse is a fictional structure made out of code. And code is just a form of speech—the form that computers understand."

NEAL STEPHENSON, SNOW CRASH

CONTENTS

ABBREVIATIONS

- 2D: two dimensional
- 3D: three dimensional
- AI: artificial intelligence
- AR: augmented reality
- DAO: decentralized autonomous organization
- DeFi: decentralized finance
- DEX: decentralized exchange
- ML: machine learning
- MR: mixed reality
- NFT: non-fungible token
- USD: United States dollar
- VR: virtual reality
- XR: extended reality

*All dollar figures refer to USD
**Crypto ticker symbols are denoted with '$' eg '$BTC' for Bitcoin

INTRODUCTION TO THE METAVERSE

We are on the edge of a new milestone in technological advancement that will mirror the Industrial Revolution of the 19th century. But this time, it has one name —Metaverse.

For the past 20 years, we have witnessed digital technology permeate every aspect of our lives. We saw 3D virtual reality (VR) tech and experienced machine learning and artificial intelligence. All this, alongside Web 3.0, has collectively enabled the creation of the open Metaverse.

When we look towards the future, there is no denying that most of life's aspects today will transition to the digital domain. We spend most of our time online —shopping, communicating, learning, working, having fun, and connecting. We have already plugged into that persistent, high-fidelity virtual realm where all people can play, build, and monetize their virtual experiences.

Naturally, our innate drive is to strive for more,

expect better experiences, and create more meaningful connections. The transition toward a digital-first environment began well before the Covid pandemic. Today, countries and businesses still deal with the effects of prolonged lockdowns and economic turmoil as we adjust to the new normal.

Business and personal interactions have slowly been delegated to the digital domain. As a result, a virtual space for companies is now the norm for those that want to enable remote working. And some of the world's biggest brands are taking note: Shopify, Netflix, and Dropbox, to name a few. Humanity is ready for the next generation of the internet or Web 3.0, which will allow people to deliver connected, immersive experiences—from social audio to virtual reality—based on real-time activities. So in a way, it is already here and improving.

When Zuckerberg told The Verge that he wanted to "build the next set of computing platforms and experiences across that in a way that's more natural and lets us feel more present with people," we already knew that the Metaverse would not only be "some kind of a hybrid between the social platforms that we see today but an environment where you're embodied in it." We need the Metaverse, so the distinction between real and virtual can truly disappear.

What happens in the virtual space has never been more real. Our relationships, the emotions we experience, the way we express ourselves. On par with everything we do and experience, we also need to elevate our sense of the digital self. And this is where the Metaverse comes in.

The Metaverse will be disruptive to every aspect of our social lives. As the digital domain is already teeming with life and loads of creative people are coming together to be a part of its growth, we are about to see some outstanding digital architecture on Web 3.0. Perhaps a world like the one we played in the popular Sims game series will come to life in a fully realized and immersive Metaverse, built on the foundations of the blockchain we see today.

Now that the pandemic accelerated the adoption of various technologies, the shift to an online world will continue. Many people now spend even more of their lives online, from socializing to working, from education to entertainment. The Metaverse takes VR upon a new frontier where it is centered on human connection.

In the years to come, we are likely to see the Metaverse dramatically affect every aspect of our lives. What remains to be seen is whether this new technology will bridge the gap between virtual and physical reality, serving the ultimate vision of enhancing humanity via digital technology without erasing privacy and individuality. So brace yourself, and get ready to enter the Metaverse.

CHAPTER 1
WHAT IS THE METAVERSE?

THE METAVERSE EXPLAINED

IN 2021, no term saw a more meteoric rise than the "Metaverse." First used in a book written by author Neal Stephenson and published in 1992, the Metaverse has changed how we perceive virtual reality (VR). From the blockchain space to fintech, corporations, start-ups, and governments are looking for ways to get on board. Facebook has even renamed its company name to Meta to better reflect its focus on Metaverse. But what is the Metaverse, exactly?

In its simplest form, the Metaverse is an online world that often incorporates VR, augmented reality (AR), and mixed reality (MR) elements. In other words, it is an amalgam of VR and AR-based experiences. As the online world evolves, the way people engage and communicate in the digital domain will change dramatically. The digitization of everything will fuel creating a new world where people can move in and out effortlessly. We see it as an extension of current digital platforms. It will, however, not replace the Internet but instead expand on top of it.

We can also understand what the Metaverse is when we look at its name. "Meta" means "beyond," and "verse" comes from the word "universe." So the word Metaverse essentially means a world beyond our current universe and into VR. But what does that mean for the world, and what are its effects on the Internet, blockchain, and NFTs?

Most people travel from home to the office and back, unaware that digital content has permeated every aspect of our lives. You've seen:

- cool menus at restaurants that move and show you tasty meals making you hungry
- face-recognition software integrated with your social media channels for more secure access
- interactive billboards that display a highly-personalized digital message to passers-by
- multitouch kiosks at the museum or bank for your convenience

The digital domain is all around us. But we are still fairly limited in how we can interact with it.

The Metaverse creates a space where humans can

participate in a shared virtual environment to continue their physical reality. The term originated from the novel "Snow Crash," written by Neal Stephenson, which described humans interacting as avatars in a 3D space. We then saw Steven Spielberg shed more light on it in Ready Player One. In this film, people can go inside a virtual world and become a 3D being or an avatar. As a result, the players experience things they can't experience in the real world, such as performing extreme physical feats or driving futuristic vehicles.

It's hard to believe that it's been more than ten years since Steven Spielberg released the film Minority Report, which was based on a short story by Philip K. Dick. His movie introduced movie audiences to a special effect-laden futuristic world. But now, it seems, most of the technology we first saw in the film is revolutionizing retail and other aspects of our present world.

The futuristic thriller "Minority Report" captivated audiences with scenes featuring gesture-controlled computer interfaces, ubiquitous digital displays, and a reality where technological advancements make the presence of media nearly boundless. Most of these concepts were used as reference points for the future of computing. They became the basis for the outbreak and personalization of digital content.

Soon after the release of this movie, brands like Google and Meta discovered the benefits of facial recognition software to deliver customized advertising. And companies in the food, service and retail industries are not far behind with interactive billboards that display a message targeted to consumers based on gender and mood in real-time. Given that the digital signage software that makes such fine-tuning possible continues to undergo development—it's only a matter of years before businesses can take advantage of everything it offers.

The Metaverse enables us to take the next step in the next evolution of social connection. It is also about creating a world beyond our current physical reality. Naturally, this also raises many questions about the dependence of humans on technology and our notion of the real world.

Many gaming studios and digital brands are already looking for ways to use the Metaverse to increase engagement and socialization in the digital domain. Some even view the Metaverse to merge the physical and virtual worlds into one digital-physical space where humans can exist as avatars. Examples of activities that people might do in the Metaverse include:

- talking to one another
- hanging out in a virtual neighborhood
- playing games and earning money
- watching movies and tv series
- browsing the web
- shopping for clothes and groceries
- meeting for work in a virtual conference room

It will be a new Internet. Instead of the current one-dimensional World Wide Web that people surf through every day, the Metaverse will have texture, dimension, and color. As a result, people will meet, watch shows, hang out, visit virtual museums and events, visit new places, enjoy new experiences, and do anything they'd like without physically leaving their homes.

We cannot fully comprehend the functionality of the Metaverse without touching upon another new technology—blockchain. Blockchain is the basis for creating non-fungible tokens that have made it possible to own digital assets, but more on that later.

THE FOUR MOST IMPORTANT FEATURES OF THE METAVERSE

The Metaverse will truly revolutionize how we connect. Yet, many people still do not know what defines the

Metaverse despite the current events. So here are the top four most critical features of the Metaverse.

Virtual world

The main element of the Metaverse is its virtual world. To truly experience a realistic virtual world, you will need access to gaming consoles, mobile devices, and wearable technology. Unfortunately, it will be almost impossible to feel present within the Metaverse without this equipment. Yet, I am confident that machine learning and artificial technology will continue to influence how we interact with the virtual world in the future.

People

The Metaverse is a social environment. There will likely be tens or hundreds and thousands of people within the Metaverse at one time, all of whom are represented by avatars. These avatars could be robots, virtual specialists, or artificial intelligence forms. You can socialize with the other members, compete against them for prizes, or work together on building a solution. In addition, Metaverse experts envision that communication within the Metaverse will be more natural than video conferencing since you have the opportunity to see eye-to-eye, which is not currently possible with video calls.

Perseverance

Perseverance refers to accessing the virtual world whenever and wherever you want. Your access will never be restricted by larger corporations or government bodies. Instead, you will be able to access it from any physical location (assuming you have a stable Internet connection). And just as social media relies on user-generated content today, so will the Metaverse.

Real-world connection

In certain Metaverse views, objects in the physical world could represent objects in the virtual world. So, for example, you could pilot a drone in the physical world and, at the same time, control a drone in the virtual world. People have referenced the combination of physical and virtual worlds as "digital twins."

Now that you understand the four most important features of the Metaverse, it's time to understand why we need it.

BENEFITS OF THE METAVERSE

Gaming is the most common way for individuals to learn about Metaverse's benefits. However, there are many other uses for it. For example, Marshmello, Travis Scott, and Ariana Grande have hosted virtual concerts, attracting millions of people in Minecraft, Fortnite, and Roblox. For content, Minecraft is a sandbox video game where users can adventure, build, and battle one another. Fortnite is a free-to-play Battle Royale game with many different game modes. But what is Roblox?

Roblox is a virtual environment fairly developed and remarkably similar to the Metaverse. Roblox is a game platform—or, to put it another way, an online amusement arcade—where you have your own unique and distinct identity for all of the different games. Roblox allows you to build, interact and socialize with your friends online from anywhere around the globe. Although the aesthetic style is a little blocky or Lego-like, the in-game physics are accurate, and the creative potential is limitless. We've all heard of Grand Theft Auto, a well-known open-world game. Consider a virtual reality (VR) upgraded version—this is the Roblox Jailbreak.

With the advancement of Metaverse devices and an ever-growing user population, it doesn't seem absurd to take virtual concerts and festivals to a new level, with more interactions and a better user experience. After all,

Roblox's CEO, Dave Baszucki, has stated that user-generated content is their priority—the Metaverse will be produced by people, not by his team.

Today, work is a necessary component of our social life. However, the way we work has changed substantially due to the COVID-19 outbreak. In July 2020, Microsoft Teams released "together mode," which may bring meeting participants into a single virtual room, allowing the speaker to connect with the audience more readily. But Microsoft has a greater goal: Mesh will allow users to engage digitally and with virtual objects, giving all meeting participants a sense of presence. Within Mesh, businesses can create virtual worlds, and meeting attendees can enter the Metaverse with their avatars and cooperate in a mixed reality environment. In addition, in Mesh, presenters can draw graphs and show their models or prototypes.

Meta did something similar at the same time. The basic service of Facebook, Inc. remained the same after it changed its name to Meta Platforms, Inc.—connecting people. Horizon Workrooms is Meta's solution for work collaboration. You can engage with all of your colleagues in the Metaverse if you log in with a work account. In Workrooms, you can even transport your desk and keyboard into the Metaverse and pin photos directly from your computer on the virtual whiteboard. Imagine learning new abilities and accumulating knowledge when the Metaverse is fully established. Students would benefit greatly from learning about sophisticated technology, fragile artifacts, or difficult subjects in the Metaverse before experiencing them in reality.

Apart from providing an immersive experience, the Metaverse can have a huge economic influence. The play-to-earn gaming paradigm heavily relies on payer involvement for adoption. It's not a novel business concept for professional gamers to sell in-game things like character skins for virtual currency or even real money. Apart from commerce in the Metaverse, we may anticipate a surge in digital art tokenization into non-fungible tokens (NFTs) and virtual art auctions. People can now acquire virtual property and land in the Meta-verse, making it a "real" business.

The following connected industries, I believe, will be most beneficial to the development of the Metaverse. Similarly, the Metaverse will aid the growth of these industries.

Metaverse Platforms

Metaverse platforms serve as the foundation for virtual environments and digital worlds where anybody can connect and engage. Meta, Microsoft, and Roblox all have distinct perspectives on the future of the Metaverse. In addition, Decentraland, The Sandbox, and RFOX are building toward the blockchain Metaverse.

Metaverse Devices

Extended reality (XR) companies create gadgets that provide access to the Metaverse. The connection between VR equipment and the Metaverse is self-evident. Consumers cannot have an immersive experience without devices. Augmented reality (AR) gadgets are crucial in the Metaverse, particularly for business collaboration. Sensors, smart glasses, haptic motors, and other interactive solutions are among the components of XR gadgets. Given that large portions of the Metaverse must be displayed in 3D, high demand for graphics processing units is unavoidable.

Metaverse Software

The foundation for an immersive experience in the Metaverse is software technology. Software technology encompasses, among other things, software frameworks or game engines, physics engines, real-time 3D content solutions, vision solutions, XR service, interactive sound systems, and video compression. Additionally, blockchain technology can revolutionize the Metaverse by offering enhanced security, greater transparency, and increased efficiency of transactions between people. But, more on that later.

IS THE METAVERSE JUST ANOTHER BUZZWORD FOR THE INTERNET?

The internet is an ecosystem of computers, servers, and other electronic devices that connects billions of people worldwide. Internet users can chat with one another, see and engage with websites, and purchase and sell products and services once they are online. For some progressive minds, the Metaverse is the next step in the evolution of the internet. Gaming, online communities, and business meetings where people participate via a digital replica or avatar of themselves are just a few examples.

Rather than competing, the Metaverse complements the internet. Users in the Metaverse employ virtual reality (VR), augmented reality (AR), artificial intelligence (AI), social media, and digital currency to navigate a virtual world that mimics aspects of the physical world. The current state of the internet is where people go to "surf" or "browse." People can, nevertheless, "live" in the Metaverse to some extent.

Governments may have the potential to expand their reach into the Metaverse. Barbados, for example, aims to create a diplomatic office in the Metaverse— particularly, the online environment Decentraland. In contrast, most governments maintain a rather static presence on the internet.

Many services have sprung up due to the internet's expansion, paving the path for creating the Metaverse. Now it's just a matter of deciding what shape the Metaverse will take. Will it be open to the public in the same way as the internet is? Or will it be more of a controlled experience by a few large corporations?

The Metaverse surely won't replace real, in-person interactions. However, it will augment every experience we, as humans, can have in the digital domain. It will create many exciting opportunities for consumers and brands alike. Whether it's large tech players such as Microsoft and Meta planning to create new workspaces, popular brands like McDonald's' opening' restaurants inside it where people can order, or music artists shifting to a digital concert format, the opportunities presented by interactive, digital worlds seem limitless. The Metaverse will likely infiltrate every sector in the coming years.

CHAPTER 2

HISTORY OF THE METAVERSE

THE ORIGIN OF THE METAVERSE

THE METAVERSE IS CURRENTLY TRANSFORMING into a decentralized network of virtual worlds. Non-fungible tokens and cryptocurrencies are freely transferred between participants in the decentralized Metaverse. Users are granted free reign over the building blocks of a frontier owned by no one and everyone at the same time. Individuals have more ownership in these parallel spaces thanks to the decentralized Metaverse. But, the Metaverse isn't a brand-new concept.

The concept of an immersive digital reality separate from the physical world may be traced to 1980s video games when the term was originally popularized in the early 1990s. The internet and its futuristic successor—an aggregation of all the digital worlds and objects produced over the decades—have been imagined by individuals in the technology community since the late 1970s. The Metaverse is finally growing into its own in 2022, with Meta and the development of blockchain technology sitting at the forefront of how it will be shaped.

SNOW CRASH BY NEAL STEVENSON

Metaverse combines the Greek word Meta and the English word universe. Meta is commonly used as a prefix in Greek to denote "beyond" and convey a meaning of extra depth to our existing reality or universe. Metadata and metaphysics, for example, allude to something other than data or physics and typically have a self-referential connotation.

The Metaverse, like its forerunners, rose to prominence amid a flurry of scientific advancements in the twentieth century. In Neal Stephenson's 1992 sci-fi novel Snow Crash, he popularized the term "Metaverse." Snow Crash would become a touchstone for some of Silicon Valley's most prominent founders in the decade. Likewise, the novel's prescient vision of a virtual universe merging with the actual one serves as a future roadmap.

The Metaverse is gradually gaining traction in public, as the notion—which was previously considered science fiction—appears to be becoming more feasible and more likely to change the way we purchase, socialize, learn, work, and play. Hiro, the protagonist of the story, travels through the Metaverse, which appears to be a small-scale urban area generated using code where people can have lifelike experiences. Many of Stephenson's Metaverse's characteristics are reflected in today's terminology:

- The Metaverse is a three-dimensional environment
- The Metaverse is a simplification of reality
- Like today's VR headsets, users can use goggles to access the Metaverse
- Users get a first-person view of the Metaverse, and they can customize virtual avatars to some extent

Snow Crash was released just one year after the official launch of the internet in 1991. While the term "Metaverse" was limited to the technical community for the first few years, advances over the next few years lay the groundwork for what the Metaverse will become in 2022 and beyond. Blockchain development in theoretical talks and primitive conceptualizations of viable artificial intelligence occurred in the 1990s. Throughout the 2000s, technology advanced at a breakneck pace, giving birth to digital twin technology in 2002. In this phenomenon, an asset simultaneously has its place in the physical and digital world. After Stephenson's successful novel, here comes another powerful influencer—this time, in 2018.

READY PLAYER ONE BY STEVEN SPIELBERG

In Steven Spielberg's 2018 film, Ready Player One, the Metaverse is depicted in a more modern pop-cultural environment. The film portrays the Metaverse in a more contemporary light, showing life in the virtual world as more engaging than physical.

Since its release in 2018, Ready Player One has sparked a flurry of conversation regarding humanity's new future, albeit in a more catastrophic context. First, however, futurists specify the number of key aspects of the Metaverse from a theoretical standpoint.

While the plot is set in the strife-torn meatspace of 2045, most of the action takes place in the OASIS (Ontologically Anthropocentric Sensory Immersive Simulation), a huge network of artificial worlds. The OASIS has become the goal for real-world virtual reality (VR) creators, many actively seeking to recreate its promise in the spirit of reality, catching up to sci-fi.

The futuristic novel Ready Player One (2011) by Ernest Cline depicts the original dystopian future in which young people, known as "the missing millions," spend most of their time in a Metaverse known as "The Oasis," which also happens to be the world's most valuable corporation. Consider what would happen if Google owned not only a search engine but also every website you visited. When James Halliday, the Oasis' founder, dies, he offers his Metaverse to anyone who can solve a 1980s trivia question.

The OASIS has one appealing feature—diversity. Users develop some of the environments in the OASIS, while government agencies create others; they range from instructional to recreational (reconstructions of 1980s fantasy novels are popular), nonprofit to commercial.

Today's multiuser VR experiences are more PUDDLE (Provisionally Usable Demonstration of

Dazzling Lucid Environments) than OASIS. Some of the restrictions are purely aesthetic: Users in AltspaceVR are limited to a small number of expressionless human and robot avatars, but Against Gravity's Rec Room's silly up-with-people charm relies on you not caring that avatars lack noses.

Other limitations are based on my experience—you can only hang out with folks you're already friends with on Meta's Spaces. Other concerns hamstring startups with OASIS-scale ambitions, whether it's a noob-unfriendly world-building system (Sansar) or a dark-side-of-Reddit feel that attracts trollery (VRchat).

The issue isn't with metaphorical boundaries; it's with literal ones. None of these PUDDLES are in contact. You can't go back and forth between Rec Room and VRchat; you're locked where you started. That is why it is not easy to feel completely absorbed. To get to Cline's 2045, developers must begin constructing the groundwork for an infrastructure that immediately connects these worlds. That doesn't sound idealistic or even hazardous.

Consider the days before the internet, when different institutions operated their private networks. A single network was only viable after computer scientists collaborated to standardize protocols. Imagine that concept being applied to virtual reality—a Metaverse in which users can flit between domains without losing their identities or bearings.

The OASIS operates because it appears to have no owners and no urgent requirements. It's a service, a toolset that both individuals and businesses can use. If we want to fulfill this potential—universal freedom and possibility—we need to look at VR the way Cline does: as an internet unto itself, not as a first-to-market product.

THE METAVERSE AND THE SIMS

Lao Mei created The Sims in 2000. In this game, the player acts as a "god" who regulates the world. This game is now in its fourth generation. You can move people, construct houses, develop land, and operate independently. You can play in different ways, and you can even make movies. Simulating the world allows you to grasp what you can't do in real life. Considering how popular Metaverse is growing, with corporations like Meta (formerly known as Facebook) rebranding and the nonfungible token (NFT) trend, The Sims appears relatively similar to the Metaverse.

When it comes to gaming mechanics, concepts such as virtual sociability, working in the Metaverse to afford products, and constructing and furnishing digital real estate are already in place. In addition, some fan-made things are frequently put into the game, laying the groundwork for collectors who want unique, limited-edition in-game items or NFTs.

This is where the concept of digital copyright mentioned by Meta comes in. Think back to the beloved Sims game franchise, where you could build houses and towns in a virtual environment. The basic package with in-game items was offered for free by EA Games. However, as more people started playing the real-life simulator, their creators started selling new creations online—luxurious villas, new skins, custom hair and eye colors, designer items, and furniture.

The Metaverse takes the next step in digitizing this virtual environment by making it more relatable to people. It's more about experiencing the world via your avatar rather than passively observing a character in a game. In a way, The Sims has prepared us for the Metaverse. However, we are yet to experience the full extent of the engagement and interaction it can offer.

THE FACEBOOK METAVERSE ANNOUNCEMENT

Facebook intends to focus on the future, changing its name to Meta. Following weeks of intense scrutiny after releasing thousands of internal research documents and messages to regulators, lawmakers, and the press by a whistleblower, Facebook attempted to turn the page by emphasizing its futuristic technologies.

According to Mark Zuckerberg, Facebook's CEO, the future of technology is a digital world where people can communicate with one another. These parallel universes, which he refers to as the "Metaverse," will enable new forms of art, music, entertainment, and commerce. But, according to the corporation, it will also be created ethically, with privacy and interoperability in mind.

Zuckerberg announced an ambitious new effort to

his staff in June 2021. The company's future would go beyond its mission of developing a collection of connected social apps and supporting hardware. Instead, Zuckerberg stated that Facebook would create a maximalist, interconnected set of experiences right out of science fiction's handbook, dubbed the Metaverse.

In a presentation to staff, the CEO said the company's divisions focused on products and solutions for communities, artists, commerce, and virtual reality would come together and cooperate to accomplish this ambition. "What I think is most interesting is how these themes will come together into a bigger idea," Zuckerberg said. "Our overarching goal across all of these initiatives is to help bring the Metaverse to life."

"If we all work at it, and within the next decade, the Metaverse will reach a billion people, host hundreds of billions of dollars of digital commerce, and support jobs for millions of creators and developers," Zuckerberg said during a presentation at one of Facebook's conferences. "We are fully committed to this."

Facebook's Metaverse efforts are still in their early stages. However, the corporation has released several software tools, including technology that allows users to operate programs with their voices or hands. Facebook's hardware products are also growing. "This is the next chapter of our work, and, we believe, for the internet overall," Zuckerberg said.

Meta's efforts came when the corporation dealt with one of its most contentious issues. Following The Wall Street Journal's series, The Facebook Files, more than a dozen news organizations have published articles based on the trove of hacked Facebook documents during the last week. The internal study, memoranda, and argument were cited in the publications, highlighting the firm's difficulties in policing its social networks worldwide.

Zuckerberg responded to the disclosed documents, stating that the firm is attempting to balance free expression and limit harmful content. According to him, Facebook employs more than 40,000 people who work on safety and security. In addition, the social media giant is expected to spend more than $5 billion on the topic in 2021. However, content moderation in virtual reality has its own set of issues.

Zuckerberg outlined his idea for how we'll work and play in the Metaverse at Facebook's annual virtual reality conference on Thursday. He also talked about Cambria, the company's next-generation virtual reality headset. Like other VR headsets, Cambria puts a screen so close to your eyes that you believe you're actually in the virtual world. In addition, however, it detects and transmits emotion from the wearer's face into the "Metaverse."

The company's new name, Meta, is the major news from Facebook's announcements. Zuckerberg explained that the new name refers not just to the Metaverse he intends to create but also to the Greek word "meta," which means "beyond."

"It represents the fact that there is always more to build," remarked Zuckerberg. "The story will always have the next chapter."

"The Metaverse is the next step in the evolution of social interaction. Our company's mission is to help bring the Metaverse to life; therefore, we're changing our name to reflect our dedication to the future."

THE EVOLUTION OF THE METAVERSE

The term "Metaverse" can have several meanings, although the word "meta" means "beyond," and "verse" means "universe." For simplicity, the term Metaverse refers to a virtual area that users can use to establish a new Internet that employs a three-dimensional (3D) virtual and augmented spectrum. Many omnipresent technologies are involved in Metaverse innovations, including blockchain technology, virtual reality (VR), augmented reality (AR), 3D reconstruction, artificial intelligence (AI), machine learning (ML), and the internet of things (IoT).

The Metaverse has been described as a universe where everything interacts digitally except you. Land, avatars, buildings, art, and even names, for example, can all be bought, sold, and stored on the blockchain. It won't be long before Metaverse blockchain technology enables us to explore various locations where you can meet with others, attend events, tour lands, and even purchase goods and services. Blockchain technology has aided in creating a magnificent virtual realm that benefits both users and businesses.

AR will power the Metaverse, with each user managing a character or avatar. For example, a typical day in the Metaverse could involve:

- waking up and saying good morning "gm" to your pet bunny in VR

- meeting with work members in a virtual workplace
- ordering food in a virtual food court and getting it delivered to you in the physical world
- finishing work and playing a blockchain-based game
- managing your finances in decentralized applications

—all from within the Metaverse.

Some characteristics of the Metaverse can already be seen in virtual video game worlds. Games like Second Life and Fortnite and work socialization tools like Gather bring various aspects of our lives together in virtual environments. These applications aren't the same as the Metaverse, but they're close. The Metaverse does not yet truly exist.

The Metaverse will incorporate economy, digital identities, decentralized government, and other applications, in addition to games and social media. Even today, user-created valued objects and currencies aid in developing a single, united Metaverse. These characteristics make blockchain a viable candidate for powering this future technology.

This type of experience, according to its proponents, promises to free us from our screens' fragmented, two-dimensional experience. Some of that liberation does point to new creative avenues. Architects, for example, may create a whole new line of work in the Metaverse by constructing virtual locations for public gatherings. In addition, fully simulated interaction—robots on space stations—will be a boon for those who have trouble engaging in traditional face-to-face discussion or who want to change their physical persona somehow.

But there's something strange about employing the vocabulary of presence to describe these kinds of immersive settings. To generate the Metaverse's "consensual hallucination," all players must be strapped into a VR headset and cut off from reality, to use a phrase from sci-fi novelist William Gibson. Even if the goggles become lighter than before and the screens become crisper than ever, it's unclear whether this is a physical experience that most people want to have regularly.

Previous interface revolutions led our machines toward more human-centric methods of conveying information—the graphic interface's visual metaphors multitouch's tactile responsiveness. However, VR requires you to establish a strange relationship with your environment with its creative potential.

Ultimately, Metaverse novels, films, and games have inspired us about the Metaverse and the seamless convergence of our physical and digital lives. Cinematography and storytelling allowed the Metaverse to take shape before our eyes, and the future is now brighter than ever.

CHAPTER 3
THE TRADITIONAL METAVERSE AND THE BLOCKCHAIN METAVERSE

THE TRADITIONAL METAVERSE

THE METAVERSE IS a virtual environment where users can create digital avatars to enable themselves to "live" in it. People can communicate with friends, acquire and sell digital assets, go to virtual locales (which might be entirely made-up or have real-life parallels), and more in the Metaverse.

The Metaverse promises a world of limitless possibilities, similar to the OASIS in Ready Player One, where the user's imagination is the only limit. The Metaverse comprises interoperable technologies like virtual reality and augmented reality. Moreover, it works on a functional digital economy powered by digital currencies or cryptocurrencies.

Furthermore, there is no single, distinct Metaverse in the current world. Instead, there are many iterations in the Metaverse. For example, if you're playing Fortnite, you can enter a Metaverse. Likewise, if you use Face-book Horizon, you can also access a different Meta-verse. The real and true definition of the Metaverse is destined to be interoperable, which means that you should be able to access assets acquired on one platform and use them on another.

In addition, you should be able to move from one area to another—even if different companies or organi-zations have built them. This concept is vastly different from the traditional Metaverse, which does not utilize blockchain technology—but more on that later.

EXAMPLES OF A TRADITIONAL METAVERSE

With the growth of the Metaverse, the possibilities are endless. Here are some real-world examples and instances of the traditional Metaverse:

Ready Player One

Ready Player One is frequently used as an example when discussing the Metaverse. There is, however, a valid reason for this. Ernest Cline's science fiction novel from 2011 presents a vivid image of what the Metaverse might look like and how it might function. In the novel, set in 2045, people turn to the OASIS, where they can interact with other players, visit different locations, play games, and even shop, to escape a world ravaged by war, poverty, and climate change.

The OASIS is a world where anything can happen—people's imaginations only constrain "reality," and anyone can be whatever they wish. Meta is well on its way to launching Horizon Worlds, its version of the OASIS in the real world, where users can explore, play, create, and engage with others in this vast digital environment.

Fortnite Concerts

What began as a game has swiftly evolved into something more complex and capable of providing a wider range of experiences. Players in Fortnite can create their worlds and embark on adventures. In addition, they can play with other Fortnite gamers in the community. The game's crossplay feature allows players to play it on various platforms, including Xbox, PC, Playstation, and mobile phones.

Furthermore, Fortnite has evolved into more than just a game, with players able to hang out and attend in-game concerts. Travis Scott, Ariana Grande, and Marshmello were among the performers who performed. In addition, Fortnite's developer, Epic Games, raises the stakes by launching the Soundwave Series, which includes music from musicians worldwide. The Series gives gamers access to in-game interactive experiences.

Second Life

Second Life is an online environment where users may create digital avatars and explore the world, connect with other users, and sell products and services using the Linden Dollar, the in-world money. Second Life is a forerunner of the traditional Metaverse. Users can interact with one another and the digital world in a shared virtual realm. It has been around since the late

2000s and allows users to explore the Metaverse's potential.

Online, multiplayer, role-playing worlds like The Sims or Second Life have been around for nearly 20 years. Modern equivalents like Minecraft, World of Warcraft, and Fortnite have hundreds of millions of users and huge supporting economies. These games were part of the digitization of our lives and normalized more persistent and multi-purpose online engagement and communication. This combination of technological, social, and economic drivers results in explosive interest in the Metaverse.

It is fantastic to see that large companies such as Meta and Microsoft are building toward their versions of the Metaverse. However, "With great power comes great responsibility." We must address difficult challenges as the Metaverse provides the opportunity to us all.

OVERCOMING UNWELCOMED BARRIERS

Data remains largely constrained by protectionism in an age where information has become as significant as any natural resource. Whether out of genuine yet misunderstood concerns for privacy and cybersecurity or simple, unabashed protectionist instincts, governments worldwide have pursued policies designed to impede cross-border data flows, localize data, and stifle digital trade. These digital trade barriers increase costs, cut against the competition (especially for small Internet-based businesses), and frustrate innovation.

As countries continue to erect digital trade barriers, the value of the global information flow diminishes. But the onset of cloud computing and electronic commerce propelled businesses to the digital domain, highlighting the need for global data distribution—freely, readily, and without regulatory restrictions.

With the possibilities of blockchain and the tokenization of assets, we see the digital domain slowly shift toward the Metaverse. A shared digital space will influence business, culture, and design, enabling new forms of engagement and experiences. The Metaverse will be the support platform for all our future online and offline interactions.

However, the Metaverse is more than cryptocurrency trade, more than social media, more than smart contracts, and more than virtual reality—t is a place where all of this meets and where the new buttons and features will be perpetually and constantly expanding. Ultimately, the Metaverse will become the bridge between business and eCommerce. Investors will trade cryptocurrencies virtually but with "real" values that can exceed hundreds of thousands of dollars.

A blockchain is a digital ledger consisting of an ever-growing list of data (or blocks) linked together using cryptography techniques. This technology creates a permanent record of transactions, usually a decentralized and public ledger. The most well-known blockchain-based cryptocurrency is Bitcoin. When you buy bitcoin, the transaction is recorded on the Bitcoin blockchain, spreading the information to thousands of computers worldwide.

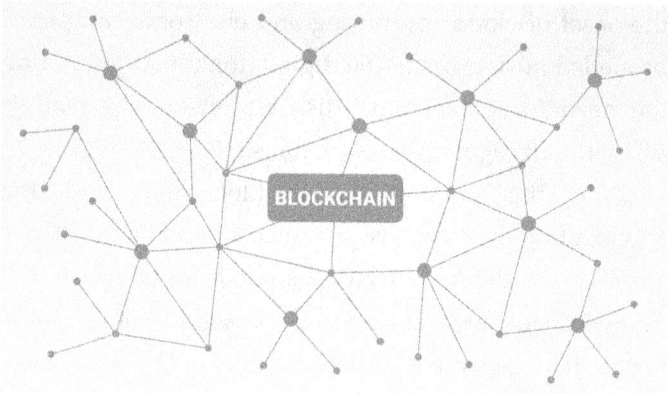

It's exceedingly tough to hack into this dispersed recording system. Moreover, in contrast to traditional banking records, public blockchains such as Bitcoin and Ethereum are transparent. As a result, all transactions are visible to anybody and everyone on the Internet. In addition, blockchain blocks will include a cryptographic hash or mathematical algorithm representing the previous block on the chain, a timestamp for the access date, and other transaction data. As a result, because there is always an end-to-end record for transparency, the blockchain is immutable and virtually impregnable to fraud. The fact that a peer-to-peer network powers blockchain ensures its security. This concept means that a block's computing power is distributed throughout a public network. As a result, each node owns a copy of the blockchain.

While blockchain offers a variety of applications (e.g., secure sharing of healthcare information, logistics monitoring, anti-money laundering, and music royalties), cryptocurrency is the most significant for the Metaverse. It is the first form of tokenization of assets that we need to see before we can transition to the virtual domain of the Metaverse, where items, experiences, music, art, and everything in between can exist in the form of NFTs and be bought and sold with cryptocurrency.

THE BLOCKCHAIN METAVERSE

Every day, in our digitally shifting world, a new technology emerges that alters our perception of reality. The phrase "Blockchain Metaverse" has recently become popular among blockchain professionals. Many sectors have adopted blockchain technology, and people are eager to learn more about it.

Several billions of dollars have been spent on digital assets and services during the last few years. That covers any digital file, data, media such as non-fungible tokens (NFTs), digital assets such as cryptocurrencies, and any other online digital service. But unfortunately, we also experienced significant dissociation from the real world due to COVID-19 and working digitally.

Blockchain technology is now used to promote the virtual world and is tightly linked to the Metaverse. Cryptocurrencies and NFTs allow consumers to purchase and sell virtual assets online. It is common for people to make money from it, and some people like creating their art as an NFT. People are also using NFTs as a type of virtual event ticket. Cryptocurrencies have been used for various purposes, including purchasing virtual land, creating NFTs, and more. These blockchain technologies form a Metaverse that can be viewed as an Internet development.

Many emerging blockchain projects use blockchain technology to construct, own, and monetize novel decentralized assets. The Metaverse was incomplete because large corporations stored everything in a centralized network before blockchain technology. Decentralization is achievable thanks to blockchain's ability to act globally as a digital source where crypto allows it.

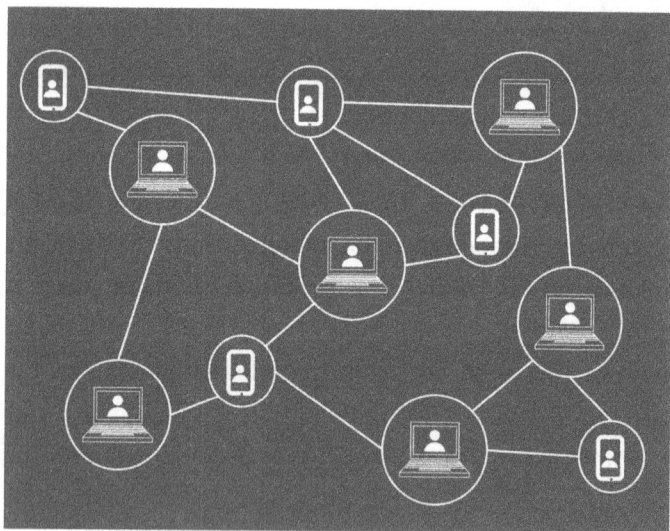

The Metaverse is not like the current version of the Internet. For example, content is available in apps and web pages on the present Internet. Conversely, the blockchain Metaverse is connected to the rest of the globe by individual nodes. As a result, there will be no need for a specific platform to access any digital domain. Every digital thing has proof of existence in the blockchain Metaverse.

BENEFITS OF A BLOCKCHAIN METAVERSE

The blockchain Metaverse is quite different from the traditional Metaverse. Here are the key differences:

Decentralization

Unlike early virtual worlds owned and controlled by corporations, a blockchain Metaverse is decentralized, with some or all components based on blockchain technology. In a blockchain Metaverse, data is stored on a digital and decentralized ledger, and owners of digital assets are genuine owners. Furthermore, assets are transferrable between applications, and the blockchain records are immutable. This structure of data decentralization also implies that individuals in the Metaverse share ownership of the Metaverse. Even if the Metaverse blockchain's original designers abandoned the project, the virtual world or game might remain indefinitely, as other designers can build on top of its existing structure.

User management

Blockchain Metaverses like Decentraland use decentralized autonomous organizations (DAOs) and governance tokens to help put their players in charge of the game's future, letting them vote on modifications and upgrades. The blockchain Metaverse can facilitate the growth of entire communities with economies and democratic government in this fashion, becoming more than just crypto games. In DAOs, the organization structure is usually flat, and all project investors have a say in the company's direction. So no more dodgy CEOs and CFOs taking charge of your money—it's all up to you now.

Verifiable provenance

In the blockchain Metaverse, in-world assets are represented by blockchain tokens, such as non-fungible tokens (NFTs). The technology introduced with NFTs modernizes in-game item standards by providing much-needed openness and access to asset markets. In addition, Metaverse tokens and other assets are coded to help authenticate the provenance of in-game user-generated material. For example, gamers might transfer an item "minted" (created on the blockchain) in a game to another game. Because of blockchain technology, this new game can verify that the digital asset is legitimate.

Economic worth in the real world

Blockchain Metaverse economies are inextricably linked to the broader crypto-economy since they utilize blockchain tokens and infrastructure. Holders of Metaverse currencies, avatar skins, and digital real estate can now trade them on decentralized exchanges and secondary NFT marketplaces for real-world value.

Additionally, blockchain-based play-to-earn games are becoming increasingly popular amongst gamers, but more on that later.

EXAMPLES OF A BLOCKCHAIN METAVERSE

The value proposition from blockchain technology provides sufficient evidence for the future expansion of Metaverse blockchain and crypto projects. Furthermore, for harmonizing with Metaverse's fundamental vision, blockchain offers a secure, cost-efficient, and transparent alternative. With that in mind, let's study some of the most promising blockchain Metaverse projects.

Decentraland

Decentraland, one of the first Blockchain Metaverse pioneers, is fundamentally a three-dimensional (3D) universe. In this 3D environment, users can create virtual real estate plots and engage in other activities such as hosting events, participating in social activities, and producing content.

Decentraland, in reality, was a well-known Metaverse project long before the hype surrounding the Metaverse began to build. The simple 2D game, created in 2016, has grown into one of the best Blockchain Metaverse crypto projects. $MANA, Decentraland's native Ethereum ERC-20 standard utility coin, is used to pay for goods and services within the Metaverse.

Decentraland has certain characteristics that qualify it for inclusion in a Metaverse crypto project list, such as its 3D interface. It also has in-game events, digital currencies, and social interaction components. Furthermore, Decentraland has recently become popular for the platform's virtual real estate non-fungible token (NFT), known as LAND.

Bloktopia

Bloktopia is the second notable addition to the top Metaverse blockchain and crypto initiatives. It's essentially a VR Metaverse game in which you must play in a skyscraper-filled environment. The skyscraper has 21 stories, symbolizing the total supply of Bitcoins (21 million). Its primary goal is to provide a Metaverse hub for socializing, work, events, and other activities. Bloktopia uses the Polygon blockchain to provide the four main features—studying, earning, playing, and building.

Bloktopia's four distinct blockchain-enabled functions solidify its place among today's best Metaverse blockchain initiatives. It can be a useful starting point for learning about blockchain and its role in the Metaverse. Bloktopia's native token, $BLOK, similarly uses the play-to-earn gaming mechanism. It facilitates real estate through REBLOK and provides advertising opportunities through ADBLOK. Users can also play various user-created games and content and design their own gaming spaces.

The Sandbox

The Sandbox is another notable example of the finest blockchain Metaverse crypto projects. The Sandbox is essentially a blockchain game that allows users to explore a virtual universe. NFTs, user-created environments, and other content are part of The Sandbox's virtual universe. The Sandbox has grown into a complex ecosystem where Ether and its native currency, $SAND, power the in-game economy.

Players can create their own digital identities and avatars, attaching them to a cryptocurrency wallet to manage their NFTs, $SAND tokens, and other blockchain assets. Then, using powerful software programs, players can construct virtual experiences and games with exclusive economic potential as NFTs.

Enjin

Because of its unique features, Enjin is an essential addition to a list of Metaverse crypto initiatives. It's a blockchain platform that allows creatives to create NFTs used for in-game currencies. Enjin has successfully released software development kits (SDKs) to make Ethereum-based NFT development easier. In addition, Enjin promised to build a secure platform for minting NFTs, which has become an important part of the Metaverse.

Liquidity for NFTs is an important concept to understand. In general, NFTs suffer from illiquidity due to the need to locate a buyer for the NFTs. You can exchange your Enjin NFT into $ENJ tokens to obtain more liquid assets. Enjin, in addition to providing a liquidity exit, also promotes scarcity and digital collectability, qualifying it for use in the Metaverse.

Star Atlas

Star Atlas is the ultimate and most likely one of the most innovative Metaverse crypto initiatives. It's a new gaming Metaverse based on multiplayer video games, real-time visuals and experiences, decentralized financial technologies, and blockchain technology. Star Atlas' foundations on the Solana blockchain cross the gap between Metaverse and blockchain technology.

Users can purchase digital assets, including land, equipment, ships, and Star Atlas' gaming Metaverse personnel. Furthermore, Star Atlas has an in-game monetary system known as $POLIS, which funds many in-game procedures. Star Atlas can swiftly ascend among the ranks of the top Metaverse crypto projects in the future, thanks to its numerous unique functionalities and intriguing experiences.

RFOX

RFOX (formerly known as Redfox Labs) builds the Metaverse for everyone. CEO Ben Fairbank "... aspire[s] to be the global leader in immersive Metaverse experiences focused on retail, media, gaming, and rewards." RFOX is smaller in market capitalization when compared to the other blockchain Metaverse giants, but their potential is limitless. The RFOX ecosystem has five core pillars:

- RFOX VALT—a completely immersive shopping, retail, and entertainment experience in VR
- RFOX NFTs—an end-to-end white-label product to enable the creating, listing, and sharing of NFT collections
- RFOX GAMES—an ecosystem of

interoperable tournament-based games focused on the play-to-earn model and market, their first being KOGs (keys to other games)

- RFOX MEDIA—an online media distribution platform that focuses on rewarding content creators
- RFOX FINANCE—a platform and service to empower RFOX holders to utilize decentralized finance (DeFi) products and services to access previously untapped revenue streams

All five pillars are interoperable and click into each other to form the complete RFOX ecosystem within their Metaverse.

The $RFOX cryptocurrency is the underlying payment mechanism for RFOX products. In addition, the $VFOX cryptocurrency is the rewards-based token for interacting within the ecosystem. RFOX's core philosophy is digital inclusion, and their solution is a virtual world in the Metaverse. Keep an eye out for this one—I feel that this will become one of the largest blockchain Metaverse projects in the coming years.

CHAPTER 4

VIRTUAL, AUGMENTED, AND MIXED REALITY

THE COMPUTER-GENERATED FUTURE IS HERE

AS WE ALL KNOW, the Metaverse encompasses a wide range of technologies, including virtual reality (VR), augmented reality (AR), mixed reality (MR), blockchain, Web 3.0, digital assets, artificial intelligence, and much more. However, VR, AR, and MR are three of the most crucial parts of the Metaverse, offering consumers a 3D immersive virtual experience. As development in the Metaverse continues to grow rapidly, it's important to understand what each of these fundamental technologies involves. So, let's look at what they are.

VIRTUAL REALITY

Virtual reality (VR) is perhaps one of the most well-known technologies in the Metaverse. VR refers to a three-dimensional (3D) and computer-generated virtual environment created by realistic sounds, images, and other sensations. Devices such as VR headsets, helmets, gloves, and body detectors are equipped with sensory detectors to replicate a real environment or an imaginary world.

Using these hardware pieces, we can immerse ourselves in the virtual environment and interact with it. However, because it does not consider our physical surroundings, it's common for users to utilize VR technology in a large space to avoid accidentally injuring themselves by bumping into things in the physical world.

Some argue that the immersive aspect of virtual reality might be isolating for the users and those around them. On the other side, virtual reality can bring individuals closer together by reducing distance, even between people across the globe (at least virtually).

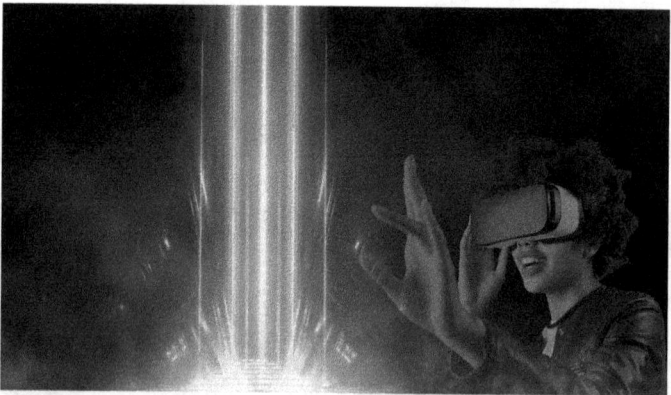

However, is VR believable and immersive? Absolutely—it's even possible to have virtual sex! Still, there are some important limitations of VR to recognize:

- some people get locomotion sickness
- there is a lack of physical world vision
- VR promotes unnatural head movements

VR and Metaverse immersion will undoubtedly continue to improve with time. However, it's promising to know that we are only halfway through Metaverse's first and second stages, and there's still plenty more to come.

AUGMENTED REALITY

The future Metaverse will likely be an augmented reality (AR) environment accessible through a see-through lens. Even though full virtual reality (VR) technology might provide substantially higher fidelity, AR will be integral to the future Metaverse. The truth is that visual fidelity will not be the deciding factor in widespread adoption. Instead, technology providing the most natural experience to our perceptual system will drive adoption. Integrating digital content directly into our physical surroundings is the most natural approach. So it's important to understand AR and how it will affect the Metaverse.

Of course, a certain level of fidelity is essential, but

perceptual consistency is considerably more crucial. All sensory information, such as sight, hearing, touch, and motion, feeds a single conceptual picture of the world within your brain. With low visual fidelity via AR, we can achieve perceptual consistency as long as virtual objects are registered spatially and temporally to our surroundings.

However, establishing a unified sensory picture of a VR environment is difficult. This idea may seem counterintuitive, given how much easier it is for VR systems to deliver high-fidelity graphics without lag or distortion. However, unless you're utilizing expensive and impractical hardware, most virtual experiences will need you to sit or stand motionless. Because of this discrepancy, your brain is forced to create and maintain two unique models of your environment: your real surroundings and the virtual world you see via your headset or glasses.

In the end, AR will rule the world. It will likely surpass VR as our primary portal to the Metaverse and the current ecosystem of phones and computers as our primary interface to digital material. After all, strolling down the street with your neck bowed and a phone in your hand is not natural for the human perceptual system to experience content. Nevertheless, AR software and hardware will likely overtake phones and computers in our daily lives in the next ten years.

This technology will open up incredible potential for artists, designers, entertainers, and educators, as they will be able to decorate our reality in previously unimaginable ways. AR will grant us superpowers with the flick of a finger or the blink of an eye, allowing us to change our surroundings. And it will feel quite genuine as long as designers concentrate on delivering consistent perceptual signals to our brains rather than focusing on absolute fidelity.

In terms of the future, the vision of a cartoonish avatar-filled Metaverse currently being promoted by huge platform providers is not completely true. Yes, virtual worlds for socializing will become more popular, but immersive media will not transform society. Instead, AR will be the true Metaverse, which will slowly become the central platform of our lives.

MIXED REALITY

Mixed reality (MR), also known as hybrid reality, combines the virtual and physical worlds to create a new environment where real-time interactions between virtual and physical objects are possible. MR devices constantly acquire new information happening in the surroundings, unlike VR, which immerses users in a completely virtual environment, or AR, which merely overlays digital content on top of the actual environment without considering its unique and dynamic composition. Physical-world data is extremely useful for integrating digital content into the physical environment and allowing consumers to interact with it. The virtual and physical worlds are interwoven with MR, which merges virtual and physical worlds.

MR can provide us with a 3D experience that is both immersive and interesting. These are our portals into the virtual world of the Metaverse. It creates a computer-generated virtual environment, similar to the Metaverse concept. Users can then use MR headsets, gloves, and sensors to explore it.

The MR work by industry professionals depicts an early Metaverse model. MR is already working on a digital environment with fantastic visual elements. As MR technology matures, extending the Metaverse experience to include physical simulations using MR equipment will be possible. Users will be able to interact with people worldwide by feeling, hearing, and seeing them.

VR, AR, AND MR IN THE METAVERSE

Many people think about gaming when attempting to define the Metaverse—but the ultimate goal is to transform how we use the Internet to connect. So far, the gaming industry has been the first to reap the benefits of virtual ecosystems. For example, World of Warcraft producer Activision Blizzard has made more than **$8 billion** in real-world money from their virtual world.

Other companies and brands attempt to follow the trend and find ways to use it for their own goals and gaming. For example, Decentraland, a decentralized business that uses blockchain technology, recently sold a digital plot of land for **$2.5 million** to a Canadian investment firm. This room will be used for virtual reality fashion shows and expanding e-commerce services with fashion labels. This example demonstrates how online virtual platforms may help firms discover new marketing opportunities.

We have already suggested virtual office spaces, which may take on a new look as the Metaverse evolves. Of course, we are all familiar with Zoom and Skype. This technology enables the entire team's presence in a virtual location. Even though these companies have features like image masking, which allows you to change the background of your video, the Metaverse goes even further. It provides 3D rendered avatars to represent you in virtual meetings based solely on your movements and facial expressions.

Similar use cases are already available on the market. For example, Virtuworx—a virtual and hybrid reality business—provides tailored solutions for offices, virtual meetings, and other events.

Many other industries, including advertising, tourism, education, entertainment, retail, design, and engineering, will benefit from the Metaverse. Any activity in the physical world has the potential to spread into the Metaverse. According to computer enthusiasts, the Metaverse will usher in a new phase of the Internet, dissolving barriers and allowing us to immerse ourselves in the digital world fully. VR, AR, and MR play an important part in creating the Metaverse. So the market is expected to grow exponentially in the next few years.

IS VR, AR, AND MR NECESSARY FOR THE METAVERSE?

Virtual reality (VR), augmented reality (AR), and mixed reality (MR) are all strongly tied to the future Metaverse. Virtual objects can be embedded in the physical world using AR technology. VR immerses us in a 3D virtual environment using 3D computer modeling. While wearing a VR headset or other devices isn't required in the Metaverse, experts believe VR and AR technology will become an important part of the new ecosystem.

For example, Meta's virtual world is accessed via VR headsets, smart glasses with AR technology, and limited desktop and mobile applications. The company has already disclosed that it is working on premium VR and AR headgear codenamed "Project Cambria." The device will support MR, according to Meta. It will incorporate new sensors that allow the virtual avatar to maintain eye contact and mimic human facial emotions. With improved technology, avatars will be able to use body language to portray human emotions better, giving the impression of real dialogue in virtual space.

One of the most challenging aspects of grasping the Metaverse is how it varies from today's VR. In a nutshell, VR is a part of the Metaverse, but the Metaverse is far larger. It incorporates social networking, VR, AR, MR, online gaming, digital assets, and other technological pieces. In addition, VR can enable true telepresence, which differs from the video conferencing we're used to.

The Metaverse aims to transform how we consume material from a 2D to a fully immersive, dynamic 3D environment. Furthermore, the Metaverse, as a shared virtual area, is predicted to alter how people interact with one another by integrating the virtual and actual worlds. It can, for example, dramatically improve communication with a virtual team, which is particularly important in a remote work setting. Thanks to the Metaverse, developers can also transform ordinary video chats into experiences that provide the impression of a real presence in a virtual meeting room.

Even if there are many unanswered questions, we can think of the Metaverse as a large-scale, multipurpose entity that is not limited to the VR or AR experience but is fully shown when these technologies are used. The combined VR and AR industry is estimated to reach up to **$300 billion** by 2024, according to Statista, and **$100 billion** by 2030, according to Morgan Stanley. As a result, the Metaverse has been compared to the next generation of the Internet, giving a richer user experience, with VR, AR, and MR as a gateway.

CHAPTER 5
NFTS IN THE METAVERSE

INTRODUCTION TO NFTS AND GAMING

ALTHOUGH THE BLOCKCHAIN Metaverse and games more broadly are still in their early stages of development, these new worlds will soon create many exciting social and financial possibilities. Blockchain Metaverses can provide users with innovative opportunities to play, invest, collect, connect, interact, and socialize.

Progress on the different and unique Metaverse platforms has skyrocketed over the last few years, turning the nascent blockchain gaming ecosystem into a global economic pillar. Metaverse games are designed to become a central part of the next phase of the Internet by connecting the immersive environments of different realities, the addictive playability of video games, the interactivity of social media, and the key value propositions that crypto and blockchain provide.

With each passing day, non-fungible tokens (NFTs) reach new heights of widespread popularity. But, you might be wondering—what exactly is an NFT? NFT stands for "non-fungible token" and refers to a one-of-a-kind digital asset. So, for example, collecters cannot exchange a unique artwork for another identical piece of artwork. But on the other hand, collectors can exchange "fungible tokens" such as a five-dollar bill for another five-dollar bill.

In 2021, we had several high-profile NFT sales, the most notable of which was a **$69 million** NFT—a collage of artwork by Beeple. In addition, the world is now preparing to welcome the Metaverse, with some of the top Metaverse games recently becoming popular. The Metaverse is a cluster of virtual worlds that continues to exist after you've stopped playing the game.

Most Metaverse games need NFTs, which are completely digital and one-of-a-kind. Furthermore, with Facebook's recent name change to "Meta," many businesses focus on the Metaverse. As a result, it's plausible to anticipate the emergence of top Metaverse games in the next years.

Currently, the gaming business is worth hundreds of billions of dollars and is continually expanding. But, historically, all that money went to the owners, not the players. In-game NFTs, on the other hand, provide users with the possibility to earn money through the use of unique, rare, and immutable tokens.

The NFT play-to-earn gaming business is valued at

more than **$15 billion**. In-game assets such as player skins, weapons, and characters drive play-to-earn games. They are now tradable items like any other thing in the current economy, thanks to NFTs on the blockchain. After players get these products in the game, they can sell them for real-world local currencies on NFT secondary marketplaces.

Axie Infinity in the Philippines is a perfect example. Filipinos have started making extra money by raising, battling, and trading digital creatures known as Axies. In reality, thanks to this blockchain game, some people have been able to pay their expenses and even clear their debts. To play, users only need a phone, access to the Internet, and a small amount of money to start the game.

The capacity to transport your avatar and assets seamlessly and instantly from world to world will be one of the essential characteristics of a functioning Metaverse. Blockchain could be the key to doing this. And The Sandbox, a virtual environment based on the Ethereum cryptocurrency, has a take on everlasting, player-owned tokens that could be the key to digital real estate in the Metaverse. But more on The Sandbox later.

THE IN-GAME/IN-WORLD UTILITY OF NFTS

When you purchase an NFT, you aren't buying the underlying rights to the artwork—you're buying a record on a blockchain ledger that identifies you as the owner of that record. Unless otherwise stated by the artist or company, you don't have any rights to the underlying artwork, and it's almost impossible to stop others from copying it. Fortunately, thanks to blockchain technology, your records on the blockchain are one of a kind, and the data is immutable, meaning that it cannot be changed or altered.

As is typically the case with most things digital, investors found all of the cutting-edge innovations in NFTs in the gaming industry. Many games have been developed along these lines, with blockchain ensuring the uniqueness of assets generated within the gaming environments.

For example, CryptoKitties is a game about collecting creatures. Investors can breed them together to create new creatures with a mix of "Cattributes" inherited from their "parents." Players can be guaranteed that each Kitty is a one-of-a-kind digital asset since its NFT ensures it. It cannot be duplicated, taken away, or destroyed. Furthermore, CryptoKitties with rare Cattributes sells for large sums of money. Hence, players have every reason to put in the time and effort required to play.

Axie Infinity, a slightly less cuddly counterpart of CryptoKitties, is another game with similar dynamics. Players mint digital animals called Axies, which they then train to fight other Axies. Again, players have numerous options to earn money within the game ecosystem because it is a play-to-earn game with a vibrant in-game economy. As a result, many individuals have been attracted to the game to earn additional money, similar to a second job or side-hustle.

The usage of NFTs to verify the ownership of in-game digital assets seems more obvious than merely using them to represent ownership of digital art. Games are self-contained digital environments that function according to their maker's rules. As a result, ownership disputes over in-game digital assets are quite common, especially when the item in question is rare.

When they become full-fledged conflicts, they are frequently beyond the power of traditional courts to handle, given how difficult it can be to identify the owner of the digital item. A self-contained regulatory solution to that challenge is the ability to ensure that a record of ownership of these precious assets is preserved in an immutable ledger. As such, NFTs and blockchain technology will surely power the future Metaverse.

ARE NFTS REDEFINING THE METAVERSE?

Blockchain technology and traditional Metaverse technology (VR, AR, machine learning, AI) are fantastic pieces of technology in isolation from one another. However, they'll reach their full potential when they're used together. Because both sides of technology contain a variety of traits and functions that complement one another, allowing them to converge in ways that make them greater than the sum of their parts. But what does blockchain technology provide that the traditional Metaverse is missing?

Digital assets

Cryptocurrencies like Bitcoin, Litecoin, and Ether are built on blockchains and decentralized, distributed databases protected by encryption. As a result, the future blockchain Metaverse can encourage a virtual world like "Ready Player One," where we can play, work, interact, and socialize with our friends in immersive environments without ever leaving our homes in the physical world.

It's already possible to buy virtual real estate plots in the popular blockchain project Decentraland using its native cryptocurrency $MANA—in fact, someone recently made headlines for a **$2.4 million** sale. So aside from land, we'll be able to purchase digital representations of almost anything we can buy in the real world and many things we can't! Governments are also joining the blockchain Metaverse hype, with Barbados recently opening the world's first Metaverse embassy utilizing Decentraland.

Buying digital assets will be the beginning of blockchain-based currencies in the Metaverse—decentralized finance (DeFi) is a fast-growing industry well-suited to operating within virtual worlds and environments. As a result, we can see more Metaverse-based lending, borrowing, trading, and investing.

Gaming will likely be one of the most intriguing use cases for the future blockchain Metaverse. The Sandbox is a virtual world where anyone can create games and experiences and trade digital goods and assets using the Ethereum-based blockchain. Atari and Aardman Animations, the creators of Shaun the Sheep, have already set up a shop in The Sandbox. Animoca Brands expect to see the private valuation of The Sandbox rise to **$5 billion** across the Metaverse and NFT frenzy.

We all know that blockchain gaming has already made it to recent news headlines. This gaming style incorporates online casino games and the more recent "play-to-earn" gaming paradigm. One of the most popular blockchain plat-to-earn games is Axie Infinity. Over a million daily active users train and fight digital creatures, similar to Pokemon. However, it varies from Nintendo's game because winners are given the cryptocurrency $SLP, along with the best earning around **$250** each day—a substantial sum in small countries where the game is most popular!

Oneto11, which bills itself as the world's first blockchain-based gaming environment, is another Metaverse game where players can earn cryptocurrency converted to actual money. Players compete for the platform's blockchain token ($1TO11) using their sports knowledge to compete against others. Blockchain gaming is expected to boom in the next few years if the Metaverse lives up to its hype, all thanks to the power of digital assets.

Decentralization

It's important to remember that, as with all future forecasts, much of this is simply guesswork—no one knows how the blockchain Metaverse will work just yet. People like Mark Zuckerberg, for example, have their ideas and are funding projects to make the Metaverse a reality. Will it, however, be a centralized, corporate-controlled Metaverse? Or will we wind up with something far more decentralized, similar to the blockchain concept? The ability of blockchain to enable smart contracts and decentralized autonomous organizations (DAOs) opens the door to alternate digital realities that aren't controlled by Silicon Valley behemoths. These DAOs can also be owned and regulated by the people who use them through safe voting processes and complex blockchain functions such as staking.

In reality, we're likely to see a mix of the two: companies creating and maintaining their Metaverses in which they set the rules, alongside publicly-owned decentralized blockchain Metaverses. It'll be interesting to watch how they interact—will an avatar born and leveled up in a DAO-driven public Metaverse be welcome in Zuckerberg's private Metaverse? Within the Metaverse, who will have the last say on creating fundamental societal concepts like identity and property ownership? These are all important questions, and we can be confident that blockchain technology will play a key role in determining the answers.

METAVERSE NFTS

Non-fungible tokens (NFTs) are expected to play a significant role in the Metaverse, according to several researchers. NFTs are tokens that live on a blockchain, and holders can use them to establish ownership of related digital assets. We've mostly seen them used to trade digital art. However, they may theoretically be linked to anything such as virtual avatars, game assets, and real estate (or should that be unreal estate?).

According to some, one of the most important applications of NFTs is that they can be used as access tokens into restricted areas of the blockchain Metaverse. Because NFTs are stored on a decentralized public blockchain (Ethereum is the most popular so far), establishing ownership of digital assets enables greater fluency of trust and transparency. For example, blockchain project RFOX hints that their NFT called "KOGGUs" provides access to secret locations, experiences, and easter eggs in their Metaverse called RFOX VALT.

In addition to enabling access to private spaces, NFTs will be used as rewards in many of the blockchain games in the Metaverse. This reward system is a supplement and addition to the other common form of blockchain rewards via tokens and cryptocurrencies, which are fungible and thus not unique. Some of the best blockchain games offer digital cryptocurrencies and limited NFTs as rewards to consistent players. As such, both have their place in the blockchain Metaverse.

Accessibility into the blockchain Metaverse is one of the most important use cases of NFTs that has recently attracted much attention. However, will NFTs reduce accessibility barriers into the Metaverse, or will they start the ultimate inaccessible Metaverse? Ultimately it's up to the developers of projects and members of DAOs to decide. So, if you are interested in blockchain technology and the future state of the Metaverse, keep a close eye on NFTs and how they contribute toward a brighter future for us all.

CATEGORIES OF NFTS

Non-fungible tokens (NFTs) were among the hottest topics in 2021. In recent years, unique digital assets built on the foundation of blockchain technology have grown in popularity. Hence, there has been an interest in understanding the various types of NFTs. Apart from a vision for altering traditional asset management, many are interested in the exciting economic possibilities connected with NFTs. Moreover, the progressive emergence of NFTs can provide blockchain developers and investors with various building and investing options. As a result, having a firm understanding of the various sorts of NFTs can assist you in making smarter decisions along your journey.

Different theories about the potential of NFTs and the benefits and risks connected with them have become hotly debated topics. With the help of blockchain technology, NFTs can show the real origins of an asset and help verify ownership of digital assets. In addition, the rising infrastructure and scope for innovation in blockchain technology can help NFTs find applications in various fields. Although the technological landscape is rapidly changing, here are some of the most common applications of NFTs.

Collectibles
One of the most popular applications and use cases

of NFTs are collectibles. The creation of Cryptokitties is the perfect example. Cryptokitties are surprisingly cute cats stored on the Ethereum blockchain as NFTs. Players can breed cats to unlock rare traits and then sell them on secondary marketplaces for a profit. In 2017, Cryptokitties became so popular that the Ethereum network became crowded, and transaction fees skyrocketed to **$451** from **$14** in a few weeks. NBA Top Shot is another example of companies using NFTs as collectibles. Basketball's greatest moments are stored on the Ethereum blockchain as NFTs and collectors can buy, sell, or trade them with others. The highest NBA Top Shot sale was for **$230 thousand**, and the NFT represented a video of Lebron James' 2020 NBA final dunk.

Art

Another popular possibility for NFTs is art. Artists can sell their artwork via direct sales or secondary marketplaces on the blockchain, eliminating the need to go through an intermediary. Therefore, artists can get a greater portion of sales. Unfortunately, this does mean that artists will need to find their buyers instead of relying on traditional marketing firms. However, many top artists are extremely active on social media platforms such as Twitter and Discord. They also keep up-to-date with the latest smart contract platform requirements to optimize sales via secondary marketplaces. One of the greatest pieces of art sold on the Ethereum blockchain is called "EVERYDAY'S: THE FIRST 5000 DAYS" by Beeple, which sold for **$69 million** in 2021.

Event tickets and access to private spaces

One other extremely popular use case for NFTs revolves around accessibility to functions, experiences, and locations in the Metaverse and, more broadly, Web 3.0 applications. For example, event organizers can distribute tickets on the blockchain for certain events such as music concerts or virtual conferences. People can purchase these tickets, and because NFTs are unique and verifiable, event organizers can manage the allocation of tickets more reliably and effectively. Also, blockchain technology solves the issue of ticket resale scams. It gives the ticket holders greater control of their purchases and sales via secondary marketplaces.

Media and music

Traditional music and media influences are also experimenting with NFTs. It is possible to connect music and media files to NFTs, allowing access to those files by someone who has a legitimate ownership claim. Rarible and Mintbase are two popular sites that assist musicians in minting their tracks as NFTs.

While artists benefit from direct access to their fans and new audiences, listeners benefit from a collector and scarcity standpoint. One of the main reasons for incorporating historical vinyl record features into NFT music is to give it a sense of exclusivity. In addition, the rise of music NFTs may offer promising opportunities for tackling music piracy and intermediary issues.

Gaming

The most popular NFTs represent in-game digital assets in the gaming industry. NFTs have sparked plenty of excitement among game creators. They offer the functionality of in-game digital asset ownership records, allowing in-game economies to thrive. Most importantly, NFTs in the gaming industry provide players with various benefits. For example, players can exchange NFTs for real-world local currencies. In addition, game developers receive royalties for each resale in a secondary marketplace. Keep reading to learn more about NFT games in the next chapter.

Domain names

Domain names have recently become popular due to a sizable airdrop to early adopters of the blockchain project Ethereum Name Service. Blockchain domain names are very similar to traditional Web 2.0 domains names as they are unique and used to store information. You can host a website, store backend data, or link your Ethereum address to a domain name. For example, a domain name transforms a long and complex user address into a short, unique, and memorable name.

Virtual real estate

Another growing application of NFTs in the blockchain Metaverse is for virtual real estate. Metaverse builders are embracing NFT technology to construct their virtual land ecosystems. Investors can purchase virtual land, represented as an NFT on the blockchain. Virtual real estate owners can monetize their land by building experiences or shops. They can also rent it out to other investors for passive income. One investor even paid up to **$450 thousand** for a virtual plot next to Snoop Dogg's in The Sandbox. More about virtual real estate in the next few chapters.

As blockchain and Metaverse technologies continue to develop over the next few decades, the opportunities will become limitless. And more experimental applications of NFTs will start to grow—watch out for verifying real-life government documents and logistics and supply chain management.

EXAMPLES OF NFTS

CryptoPunks

CryptoPunks are some of the most well-known NFTs. In 2017, Larva Labs created 10,000 CryptoPunk NFTs at random using an algorithm. They offered them to anyone interested who owned an Ethereum wallet. From there, CryptoPunks skyrocketed in value on the secondary market. The cheapest of all CryptoPunks is now worth over **$100,000**, while the NFT series has seen over **$1 billion** in transactions. CryptoPunks has attracted many investors, including Jay-Z, Visa, and Odell Beckham Jr.

Paris Hilton's Planet Paris is an unexpected supporter of the NFT movement. Hilton has written articles about NFTs, spoken about them on Jimmy Fallon's Tonight Show, and even produced her own NFT series. Planet Paris was created in collaboration with Blake Kathryn and contains a series of short videos that helped Hilton generate more than **$1 million** in sales.

Louis Vuitton video game

Because luxury businesses are built on exclusivity, and brand awareness, joining the NFT industry was a natural fit. However, Louis Vuitton's foray into the field of NFTs was one-of-a-kind. They created Louis: The Game, a video game where players must accomplish obstacles to find hidden NFTs.

Bored Ape Yacht Club

The Bored Ape Yacht Club, like CryptoPunks, is a collection of 10,000 algorithmically produced unique avatars. Ownership of a Bored Ape, created by Yuga Labs, entails more than possessing collectible and bragging rights. Holders can also use this NFT to join a virtual club of like-minded Apes. A set of 101 Bored Ape NFTs was recently auctioned off at Sotheby's for a stunning **$24.4 million**. In addition, the club has donated **$850,000** to the Orangutan Outreach charity.

Beeple's $69 million sale

Beeple's historic **$69 million** NFT auction is a must-see for everyone interested in NFTs. "Everydays: The First 5000 Days" is the most valuable NFT ever sold, the third most valuable artwork sold by a living artist, and the first NFT to be sold at a fine art auction house.

Jack Dorsey's first tweet

Twitter's founder Jack Dorsey published his first tweet in 2006, "just setting up my twttr." By selling his first tweet as an NFT, Twitter founder Jack Dorsey made headlines (and millions).

Origin of the Internet

In 1989, Sir Tim Berners-Lee created the World Wide

Web. He sold the original 9,555 lines of source code as an NFT for a whopping **$5.4 million** 30 years later. The NFT, which was auctioned off at Sotheby's, includes the source code for the World Wide Web and other items created by Sir Berners-Lee. Don't worry—the purchaser doesn't own the Internet. Instead, it's as if they've purchased the Declaration of Independence, claiming ownership of a key piece of human history.

$200k LeBron James dunk

Basketball cards and autographed merchandise are similar to NFTs. They're rare, collectible, and tradeable. As a result, it was only a matter of time until the sports memorabilia industry entered the NFT market. NBA Top Shot sold an NFT video footage of LeBron James dunking for **$210,000**.

A New York Times article

Journalist Kevin Roose wrote an article about NFTs for the New York Times during the peak of the NFT craze in March 2021. The piece had one unique feature: it was auctioned off as a nonfiction text. Much to Roose's astonishment, the story sold for **$560,000** at auction.

Coca-Cola NFTs

In May 2021, they held an auction for several NFT treasure boxes. It featured branded jackets, and holders can wear them virtually within Decentraland. In addition, all purchases included a Coca-Cola refrigerator that was completely stocked. In Decentraland, the company even held a 'can-top' party with music, giveaways, and a Q&A session. Over **$575,000** was raised during the 72-hour auction, entirely donated to Special Olympics International.

The numerous potential applications of NFTs in the Metaverse will transform the future. NFTs bring ownership and uniqueness to digital assets, and the Metaverse provides a digital world where anything is possible. The combination of the digital world and digital assets will change how we interact with one another. And to think that NFTs are the building blocks of the blockchain Metaverse, we are gifted with an opportunity to be part of something great.

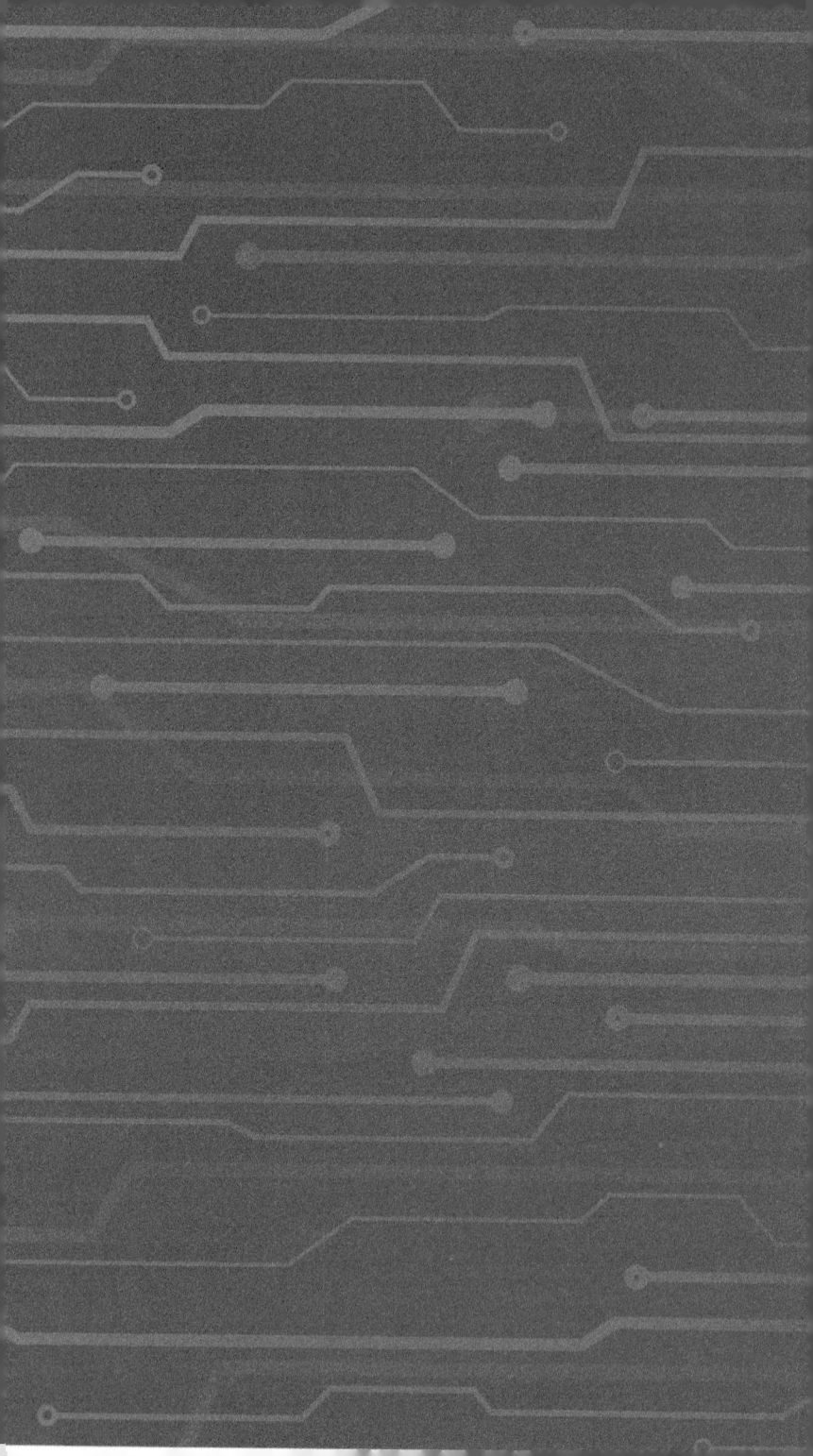

CHAPTER 6
GAMING IN THE METAVERSE

THE CURRENT LANDSCAPE WITHIN THE METAVERSE

THE TOP THREE Metaverse projects by market capitalization are related to gaming. These projects are Axie Infinity, The Sandbox, and Decentraland. Gaming will continue to be the dominating trend for most figures in the crypto industry and the Metaverse sector. The Blockchain Game Alliance's 2021 annual report indicates that non-fungible token (NFT) based games produced $2.32 billion in revenue, indicating exactly how vast the gaming-based Metaverse is.

So far, the leaders have done well enough to persuade considerable additional cash to pour in; at this point, almost any blockchain Metaverse game is getting funded without hesitation. So, there'll be no shortage of new and exciting games to play in the next few years. Other Metaverse use cases are still speculative. Companies are trying to figure out their business model and road to adoption.

Play-to-earn games in the Metaverse have pushed NFT acceptance forward, despite their opposition from traditional gamers. As a result, many in-game NFTs are available in RPGs, card games, and shooters. Weapon NFTs can aid a player in combat. At the same time, skin NFT accessories can help them carve out a unique identity in the Metaverse. And, of course, with blockchain technology, collectors have complete ownership and control of their digital assets. Many have been exchanging their unique digital assets for real-world local currencies, but more on that later.

TOP FIVE BLOCKCHAIN METAVERSE GAMES

The Sandbox

Being one of the most popular blockchain Metaverse games, The Sandbox aims to disrupt many traditionally established games such as Roblox and Minecraft. The Sandbox is unique because it gives creators genuine ownership of non-fungible tokens (NFTs). The Sandbox also incentivizes players to contribute to the game's environment and ecosystem. This unique framework is one of The Sandbox's most striking features. Players can use the VoxEdit tool to create assets and sell the NFTs that they create.

Sorare

Sorare is a football NFT blockchain game built on the Ethereum network. Almost 180 football clubs have already officially registered on Sorare, and this number is rapidly growing! On Sorare, players can purchase cards to construct fantasy teams. The cards represent real players, and the player's performances in real-world matches give collectors points. Sorare is one of the top Metaverse games to receive an almost $680 million Series B investment. Another fascinating piece of Sorare news is that Ubisoft, a well-known game development studio, has designed a blockchain-based game using the NFTs made on its platform.

Illuvium

This blockchain project is developing an open-world role-playing game built on the Ethereum network. The game's basic premise is to explore the large virtual environment. The game's primary goal is for players to develop strong beings known as "Illuvials." Illuvium, which is set to launch in the first quarter of 2022, has already amassed a **$1 billion** market valuation from early investors. The majority of Illuvium conversations feature videos from the actual gameplay experience. Illuvium's major strengths, according to Metaverse fans, are its amazing graphics and engaging gameplay.

Ultra

Ultra is the first blockchain ecosystem to power an entertainment platform that includes blockchain-enabled services from various games. Gamers can discover, buy, and play games via the Ultra platform. Also, it's a great social network for gamers to watch and support each other via streaming services, compete in competitions, and participate in contests. The network's ability to scale effortlessly above 12,000 transactions per second is one of Ultra's most noticeable features.

Axie Infinity

Axie Infinity is another popular Metaverse blockchain game. The game was first released in 2018 and has recently gained mainstream popularity. During the COVID-19 outbreak in the Philippines, Axie Infinity proved to be a lifesaver for many unemployed people. Gamers collected NFTs, which they then traded for real-world local currencies. With a market valuation of **$3 billion**, Axie Infinity is a leading blockchain Metaverse game, but more about Axie Infinity later.

WHAT ARE PLAY-TO-EARN GAMES?

For the past 50 years, home video games have been used to distract us from a hard day's work. Elite gamers have the opportunity to monetize their games via streams, sponsorships, and tournaments. However, less than 1% of us fit this criterion. But, with the birth of blockchain technology and the continued hype around play-to-earn games, many of us now have the opportunity to make passive income by playing games in our free time.

But you might be wondering what play-to-earn games are? Play-to-earn games are video games where players can earn rewards that can be exchanged for real-world value, for example, local fiat currencies. These play-to-earn games have been around for decades. But unfortunately, they often work around existing game mechanics and generally lead to scams.

With the technological advancements in blockchain technology and non-fungible tokens (NFTs), play-to-earn NFT games are becoming increasingly popular, particularly as secondary or passive jobs. For example, Axie Infinity became one of the most popular play-to-earn games in 2021, particularly in counties such as the Philippines and Indonesia. As a result, "Axie scholarships" were created. Many players in the Philippines sold their rewards to purchase real estate in the physical world for their growing families.

Because NFTs live on the blockchain, they can't be forged, and players can verify the origin of certain in-game assets. As such, there are fewer scams in blockchain play-to-earn games when compared to traditional play-to-earn games. Secondary marketplaces such as Opensea or Rarible fuel the two-sided NFT marketplace and enable trustless transactions between gamers and collectors.

Play-to-earn games are also helping with the mainstream adoption of crypto projects and blockchain technology. According to Axie Infinity's cofounder Aleksander Leonard Larsen, more than 50% of their players have never used a blockchain protocol before. However, one limitation of Axie Infinity's game is the high barrier to entry. It currently costs around **$1000** to get started, but the game designers are currently working on a solution to this problem.

Ultimately, the future of gaming is NFT games. Players will play, earn, live, and repeat this process. It could be a possible future for gamers who spend more time in video games than in the real physical world. The rise of NFTs makes this possible, and the play-to-earn model's emergence would be the final nail in the coffin.

EXAMPLES OF PLAY-TO-EARN NFT GAMES

Axie Infinity

As previously mentioned, Axie Infinity became 2021's most popular play-to-earn NFT game on Ethereum. Players can purchase cute monsters in this blockchain game and fight them against other players. Collectors can also breed Axies to make more and sell them on the marketplace for a nice profit. Additionally, players can earn $SLP tokens which gamers can exchange into local fiat currencies for real-world value. But more on Axie Infinity in the next section.

CryptoBlades

CryptoBlades is a fun blockchain game based on the role-playing game genre on Binance Smart Chain. Users can acquire $SKILL tokens by fighting in-game battles and beating foes in the game. In addition, players can design their weapons and sell in-game items on the primary or secondary marketplace. However, players must first purchase a character to begin the game. As a result, a small cost ($BNB) must be paid to begin game-play and combat fights, later refunded to the players in $SKILL tokens.

Plant vs Undead

Plant vs Undead is a popular tower defense blockchain game on Binance Smart Chain. To earn money via this play-to-earn game, you can play the survival model and earn $PVU tokens which you can sell on the marketplace. Alternatively, players can farm their land and harvest Light Energy which can also be sold or gambled away to win NFTs. If you've ever played Plants vs. Zombies, this nostalgic blockchain-based game is for you.

Farmers World

Farmers World is a popular simulation game on the WAX blockchain that has recently gained much attention. Players choose their tools, exploit various resources, build enormous farms, and live life as simple farmers. You can chip along at your farm and collect rewards. Players can then exchange these rewards for local fiat currencies.

Forest Knight

Forest Knight is a play-to-earn blockchain-based game powered by the Ethereum and Polygon blockchains. Forest Knight is a turn-based strategy game. Players need to build a team of heroes and fight against evil across the land. It involves fun and exciting PVE (player vs. environment) RPG (role-playing game) experiences, a PVP competitive landscape, and social gameplay modes. $KNIGHT is the token used for in-game utility, staking to reward loyal players, and as governance, used as a metaphorical voice in the direction of the game roadmap.

The Sandbox

Another example of a play-to-earn blockchain game is The Sandbox on Ethereum. It is one of the most popular virtual blockchain worlds. Anyone can create a platform to earn, own, and sell a virtual gaming experience. In addition, the Sandbox has a grand opening for newcomers to obtain a free NFT by registering. LAND is the Metaverse location where you can access digital content that will help you monetize your experiences, and $SAND is the token that investors may use to purchase land and other assets within the Metaverse. But more on The Sandbox in the next chapter.

AXIE INFINITY (A CASE STUDY)

Axie Infinity was founded in early 2018 by Trung Nguyen, Aleksander Larsen, and Jeffrey Zirlin. Axie Infinity was developed by Sky Mavis, a for-profit company based in Vietnam, founded shortly afterward. Axie Infinity raised around **$10 million** across various ICOs and sales of in-game assets between 2018 and 2020. Sky Mavis has also raised over **$160 million** from various VC firms since 2019. Although Axie Infinity has been live since late 2018, adoption of the game exploded earlier this year, when Sky Mavis introduced the Ronin side chain—a layer 2 scaling solution on Ethereum, specifically designed for Axie Infinity.

Axie Infinity is often described as a combination between Pokemon and Tamagotchi. The difference is that you earn cryptocurrency while you play. Every item in the game is an NFT that can be bought and sold on the Axie Infinity marketplace. Axie Infinity's marketplace has seen more than **$3.5 billion** in NFT sales, making it the second-largest NFT marketplace by trading volume after OpenSea.

Axie infinity currently offers two game modes. Adventure, where you battle monsters, and Arena, where you battle players. To play Axie Infinity, you must purchase three Axies which are NFTs. Now there are many different classes of Axies, and each class has its strengths and weaknesses when it comes to combat.

Every Axie also has unique physical traits, which gives them additional combat abilities. Axies can be bred, and the likelihood of the offspring of two Axies will express a physical trait depends on whether the trait is dominant, recessive, or minor recessive. Breeding Axies requires the Smooth Love Potion ($SLP) token and a small amount of Axie Infinity Shards ($AXS). A portion of the $AXS used for breeding Axies is sent to the community treasury and a portion of marketplace sales.

$SLP is the cryptocurrency you earn while playing Axie Infinity, whereas $AXS is the governance token for Axie Infinity's ecosystem. $AXS can also be earned when competing in battle tournaments. $AXS was the cryptocurrency sold during Axie Infinity's ICOs. It has a maximum supply of 270 million, half of which will go to the community with a five-year vesting schedule. By contrast, $SLP is a fair launch cryptocurrency because it began at 0 and was bought into existence by Axie Infinity's users. $SLP also has no maximum supply.

Limitations on the amount of $SLP players earn each day in adventure mode ensure that Axie Infinity's play-to-earn economy is sustainable. The intense competition from other players in arena mode likewise limits how much $SLP a player can earn. $SLP is also burned every time an Axie is bred. An Axie can only be bred seven times, and the amount of $SLP required to breed an Axie increases with each breed. As a result, $SLP used for breeding is burned.

Active Axie Infinity players can earn between **$400** and **$500** per month. This money might not be a lot if you live in a developed country. However, it is a lot in developing countries such as the Philippines and countries experiencing record inflation such as Venezuela. Not surprisingly, this is where most of Axie Infinity's players are based, and some of them have made life-changing gains by playing the game.

I have tried to cover the fundamentals of the project. However, please keep in mind—that these are changing all the time, and it is almost impossible to stay up to date with information from a book. So if you want to get up-to-date information on the Axie Infinity project, please check out their main website and social media channels.

CHAPTER 7
VIRTUAL REAL ESTATE IN THE METAVERSE

WHAT IS METAVERSE REAL ESTATE?

REAL ESTATES in the Metaverse are parcels of land in virtual worlds, often represented as pixels on a screen. However, they are more than just digital images —they are programmable spaces in virtual reality platforms. People can socialize, play games, sell NFTs, attend meetings, go to virtual concerts, and do countless other virtual activities in the Metaverse.

Virtual blockchain worlds (VBWs) are social platforms made up of several virtual land parcels owned by users, companies, and investors rather than a single company. Anyone can buy or rent land, allowing them to develop anything they wish, usually centered around experiences for themselves and others.

Facilities such as museums, casinos, indie games, villas, art galleries, conference centers, eCommerce stores, and many others have already been established.

But what distinguishes these facilities from standard virtual worlds such as Second Life or games like Grand Theft Auto?

Unfortunately, everything belongs to the companies that built centralized virtual worlds and games. All content is hosted on their servers, giving them complete control. When you "own" in-game assets, it's almost impossible to find real-world value outside the game's ecosystem. Generally, you will not be unable to trade or gift them to other players or take them out of the defined ecosystem.

With blockchain technology and NFTs, owers of in-game assets can be bought, sold, or traded with other players and collections. The items can also be transferred freely throughout the Metaverse and used seamlessly between applications. For example, you might buy an audio NFT from your favorite artist at a concert in the Metaverse. Then, you can submit it to a radio channel within a Metaverse game and receive royalties every time it is played to other gamers.

Digital real estate is destined to grow and expand with the rise of the Metaverse. In fact, with the Metaverse real estate boom in the last quarter of 2021, Facebook changed its name to Meta and indicated an interest in building on top of the Metaverse. As its popularity grows, the value of Metaverse real estate is forecasted to have a CAGR (compound annual growth rate) of 31.2% from 2022 to 2028.

WHY DO PEOPLE PEOPLE BUY VIRTUAL LAND?

Now that you know what virtual real estate in the blockchain Metaverse is, here are some of the most common reasons you might want to purchase virtual land for yourself.

The emergence of a new asset class

The asset class of digital real estate has established itself as credible. Moreover, its value is expanding exponentially, making it a promising investment. Similar to real-world art and real-world real estate, it appears to have the potential to become a viable financial asset.

Fear of missing out

A key driver for people to buy virtual lands is a strong sense that they might miss out on something fantastic. Many consumers missed out on buying Bitcoin at a low price, forcing them to search into alternative assets like virtual land.

Massive profits

Virtual land can offer large returns due to its alignment with the constantly increasing crypto-investment world. Many people have made thousands of dollars quickly because of the convenience of flipping land (like they do with real estate) and the volatile nature of cryptocurrencies.

Extra source of income

In the long run, virtual land opens up new possibilities for what can be done with the property, such as establishing art galleries, running marketing campaigns, or simply renting it out to others to build on and earn money. On their virtual territory, some people build virtual casinos. Big retailers are also investigating the prospect of building virtual reality storefronts.

Easier than buying a house

In addition, digital real estate avoids the key drawbacks of traditional real estate sales. These drawbacks include cumbersome paperwork, land maintenance, and tax complexities. Land purchases are also more secure and traceable because of blockchain technology usage.

Low entrance threshold

Land prices are rising worldwide, but virtual ones offer similar benefits for less than 1% of the price. As a result, real estate purchases may be beyond reach for many individuals. However, virtual land purchases have a lower barrier to entry. Therefore, younger buyers might suit this investing style as they typically have less capital than older folks.

THE TOP 3 VIRTUAL BLOCKCHAIN WORLDS

Decentraland, Somnium Space, and Cryptovoxels are the three primary virtual blockchain worlds. Because each has its unique style, you should try all three to discover which you prefer.

Decentraland

Decentraland is the largest and most popular virtual world. It is unique because it is owned and operated by its users rather than hosted on a central server. All material is hosted on the Ethereum blockchain, distributed throughout the globe, and powered by multiple computers. This idea implies it can't be turned off and will live on the blockchain indefinitely. Note that Ethereum is still using proof of work at the time of writing but is soon to switch to proof of stake.

Space Somnium

Somnium Space is the second-largest virtual blockchain planet. Somnium Space looks the most like the Ready Player One of the virtual blockchain worlds, with slick, sensual graphics and a more genuine feel. This platform, available in both 2D and VR, is the most advanced option if you want a full Virtual Reality experience.

Cryptovoxels

Cryptovoxels is the third-largest virtual blockchain world. Interestingly it is the only one constantly increasing in size with no defined limit on how big it might grow. CryptoVoxels is the easiest virtual blockchain environment to get started with and build in. All you need is a simple URL to get started if you want to learn more. No special software or hardware is required. In-world construction is accomplished by dragging and dropping pieces in real-time.

HOW TO BUY LAND IN THE METAVERSE

To purchase land in the Metaverse, you'll need some digital cryptocurrencies. Which cryptocurrency typically depends on the Metaverse or project you want to invest in. Ethereum ($ETH) is a common cryptocurrency, as are $SAND (for The Sandbox) and $MANA (for Decentraland). Finding out which blockchain, project, and land you want to invest in is the first step.

The Sandbox and Decentraland are the two most popular platforms to invest in when owning virtual land and property. However, I am sure that other projects will continue to battle these two leaders' market share in the future so make sure that you do your research before blindly investing.

You can purchase land directly from their websites using cryptocurrencies from your web wallet. Sales and ownership of Metaverse land are stored on the blockchain as NFTs, so you'll need to get comfortable interacting with Web 3.0 applications.

In addition, just like physical world real estate, you can purchase land from other owners via a marketplace. Platforms such as Opensea and Rarible act as marketplace facilitators, enabling investors to list land (and NFTs more generally) for sale and buyers to purchase land from others.

All projects are vastly different, and the landscape is changing every day. So you'll need to do your research from here for a complete, up-to-date guide on investing in The Metaverse. Start by studying the project's official website and checking multiple sources before purchasing anything.

LEARN FROM THE BEST VIRTUAL WORLD INVESTORS

When real estate prices started skyrocketing at the start of the pandemic in 2020, virtual land in the Metaverse was quick to follow. According to CNBC, interest in digital real estate prices has gone up 400% to 500% since Facebook's much-hyped transition to Meta. Also, crypto-asset manager Grayscale estimates that the digital world may soon grow into a $1 trillion business.

The digital world is slowly becoming as important

as the physical world. As the Metaverse is meant to encompass everything that exists virtually, from digital art to virtual worlds, virtual real estate parcels that are being snapped up can be seen as just one type of Metaverse investment. This phenomenon has created a new subset of investors—virtual real estate investors.

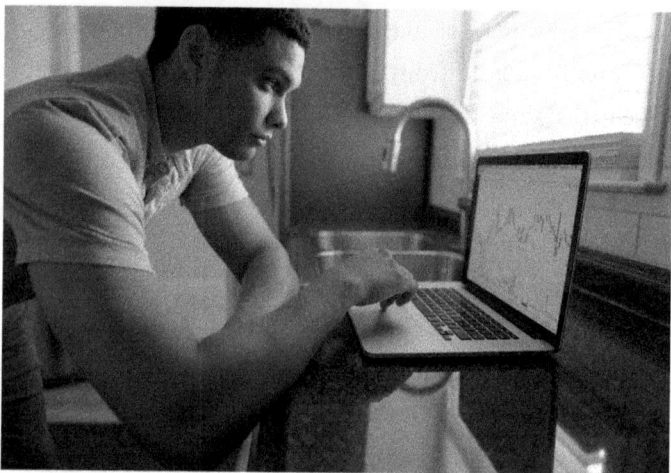

These are people or companies that purchase digital real estate. Luxury brands and higher-end retailers can establish their brands in the virtual world before entering physical retail spaces. Individuals can develop their virtual property into anything they want. If someone else buys the land surrounding theirs, they could put up shops, buildings, or even theme parks. The options are limitless.

Digital real estate is the future, and investors are already on board. Here are some of the largest and most well-known virtual-real-estate investors in the space.

Metaverse Group

Tokens.com's Metaverse Group recently purchased a piece of digital land for 618,000 $MANA or about $2.43 million. The information came from Decentraland, a virtual-only online environment where the land patch is located. The buying price quadrupled the previous virtual real estate sale record.

Republic Realm

Republic Realm, a digital real estate investment firm, reportedly paid over $1 million for an NFT that represented a plot on Decentraland. The plot included around 16 acres of digital land. The purchase price was slightly greater than Manhattan's typical home price.

Sale of Axie Infinity Land

Early Investors sold one of Genesis Land plots belonging to Axie Infinity for 550 ether ($2.3 million at the time). The previous largest transaction on the platform, which features colorful, competitive creatures known as Axies, was for 888.25 ether, or more than $1.5 million, for nine plots of land. The current boom in cryptocurrencies tied to the Metaverse, such as Decentraland's $MANA token, Sandbox's $SAND, and Gala Games' $GALA, has been fueled by the rapid activity surrounding that reality.

Virtual Yacht Sale On The Sandbox

Early investors sold a virtual mega yacht for $650,000 on The Sandbox, which isn't real estate but closely tied to the Metaverse. The yacht was created for The Fantasy Collection range of luxury NFTs intended for The Sandbox virtual game world by Metaverse developer Digital Republic.

THE SANDBOX (A CASE STUDY)

The Sandbox was founded in 2012 by Sebastian Borje and Arthur Madrid. The Sandbox was made by Pixowl, a software company based in San Fransisco. The Sandbox began its journey as a 2D mobile game where users could build their virtual worlds. The Sandbox earned interest from over 40 million users and is still available on mobile today as The Sandbox Evolution. Following the crypto bull market in 2017, Pixowl announced its plans to turn the Sandbox into a 3D game built on Ethereum where every in-game item that users create is a non-fungible token (NFT).

In 2018, Pixowl was acquired by a software company from Hong Kong, Animoca Brands, focusing on digital entertainment, blockchain technology, and gamification. As part of this acquisition, Animoca Brands partnered with TSB Gaming Limited, a for-profit software company based in Malta, which works on implementing blockchain technology into The Sandbox.

The Sandbox's development is funded by The Sandbox Foundation and employs over 100 people from 32 countries worldwide. The Sandbox first raised **$7 million** across three $SAND token sales in 2019 and 2020. Since then, they have raised millions more from various land sales in 2019.

The Sandbox's virtual world contains 166,000 plots of LAND. LAND is an ERC721 token on Ethereum, which is Ethereum's NFT standard. Investors can group these LAND tokens into estates owned by one person or districts that two or more people own. Did you know that LAND in The Sandbox is owned by various companies such as Binance, Gemini, Coin Market Cap, Atari, and even the South China Morning Post?

LAND can be customized using VoxEdit. It is also used to create in-game assets as NFTs on Ethereum. Investors can also monetize LAND by renting them out or creating pay-to-play experiences such as quests using The Sandbox Game Maker. Furthermore, LAND and assets can be bought and sold on The Sandbox's NFT marketplace, and all NFTs are priced in the $SAND token.

$SAND is an ERC20 token on Ethereum, and it is The Sandbox's in-game currency. $SAND can be staked for rewards. Both $SAND and LAND give governance rights to their holders via a decentralized autonomous organization (DAO) structure. Holders of these tokens can vote on important issues like foundation grants and feature prioritization on the platform roadmap. Holders of $SAND might vote for themselves or assign voting privileges to other players.

Sandbox has captured the attention of many players and investors as the potential is there. One research report predicts virtual gaming worlds alone could be worth **$400 billion** by 2025, with the broader Metaverse industry worth over **$1 trillion**. The Sandbox will always be one of the groundbreaking names in decentralized virtual worlds. I'm sure that it will continue to fuel the growth of virtual real-estate demand over time.

I have tried to cover the fundamentals of the project. However, please keep in mind—that these are changing all the time, and it is almost impossible to stay up to date with information from a book. So if you want to get up-to-date information on The Sandbox, please check out their main website and social media channels.

CHAPTER 8
THE NEXT BIG INVESTMENT OF A LIFETIME

CURRENT STATE OF THE METAVERSE

MOST ANALYSTS still regard the gaming environment as the Metaverse's "beginning point." However, there are already many massively popular games. Moreover, did you know that gaming is becoming a widespread consumer activity globally, with roughly 59% of Americans identifying as gamers?

Games are an excellent Metaverse environment because they already inspire us to immerse ourselves in digital worlds and join communities not limited by physical geography. In addition, consumers may currently enjoy and engage in unique experiences provided by games. This phenomenon has been increasingly widespread in recent years, as the pandemic has reduced the number of opportunities for face-to-face contacts and unique experiences.

For example, Epic Games' CEO says his firm is

investing in the Metaverse, with Travis Scott and Ariana Grande performing in their gaming settings and an immersive re-imagining of Martin Luther King Jr's historic "I have a dream" speech. Epic Games is also working on photorealistic digital humans with the Metahuman creator, showing how users may make avatars to explore the digital world more deeply.

Another example of a Metaverse embedded into a game environment is Roblox. The website, created in 2004, contains many user-generated games in which players can construct homes, act through unique scenarios, and construct residences. Following the initial public offering (IPO), Roblox CEO David Baszucki took to Twitter to express his gratitude to everyone who helped the platform go one step closer to realizing its vision of the Metaverse. Roblox has since partnered with businesses to create virtual skating play-grounds and open gardens where you can try on Gucci goods.

The number of people participating in virtual experiences is rapidly increasing. The gaming scene promotes user-generated content, virtual commodities, and locations. In addition, it provides a convenient entry point for individuals who might be hesitant to venture into the Metaverse otherwise. People who are already familiar with Minecraft, for example, are excited to attempt new virtual experiences within the game, such as concerts and events, because the setting is familiar to them.

In addition, the gaming environment provides a

perfect platform for the creation and testing of new technology, such as augmented reality (AR) and virtual reality (VR) mechanics, content moderation, cryptocurrencies, digital assets, and many more. According to surveys, participation in the Metaverse is already rising at a fantastic rate inside the gaming scene.

Sixty-five percent of individuals have participated in a media event such as seeing an in-game TV show, movie, or premiere on a gaming platform or attending a live concert.

In a gaming environment, 69% of players have engaged in social activities such as socializing, meeting new people, or traveling to digital locations.

Seventy-two percent of people have engaged in Metaverse economic activities, such as buying in-game things, investing in in-game currency, shopping at virtual markets, or trading with other gamers.

The gaming environment gives a simple entryway to various social, media, commercial, and even enterprise-level experiences in the digital era. We already have many examples of the physical and digital worlds colliding in gaming. As we invest in many "digital" events in the aftermath of the epidemic and include new technologies like extended reality into the experience, this overlap becomes ever more apparent.

FUTURE STATE OF THE METAVERSE

In many respects, the gaming industry is the Metaverse's foundation. However, the opportunities that arise with a growing Metaverse are endless—from social interactions and workplace cooperation to entertainment and the birth of a new economy. Here are just some of the potential applications within the future Metaverse.

Metaverse Social Interactions

Beyond the Metaverse, digital communities have existed for a long time—practically since the internet's inception. We've used everything from forums and social media platforms to video games to communicate with other people. Since the pandemic, social encounters have become one of the most recognizable pillars of the Metaverse. Many of us went to the digital realm to recreate human relationships during social alienation and lockdowns. Online communication with loved ones and coworkers began on various platforms, ranging from gaming environments to virtual reality hubs.

Companies like Epic Games have offered venues where individuals can hang out with their pals in "Party Royale" mode in the previous few years. Animal Crossing began hosting virtual graduations and even wedding ceremonies for its players. Even "Roblox birthday parties" were popular, allowing individuals to connect even if they weren't physically present.

From concerts to marriages, we could see a variety of festivals and events taking place in the Metaverse as we move forward. In addition, experts believe that social encounters in the Metaverse will incorporate more forms of simulated presence and human connection as technology advances.

Consider being able to holographically transport oneself to a birthday party that you couldn't make. New technologies will emerge to improve the quality of spontaneous social connections using ambient communications and lifelike human avatars to generate a greater sensation of "presence."

Metaverse Business Operations

The Metaverse's immersive, completely accessible, and open nature makes it ideal for reproducing crucial human relationships. It's worth emphasizing, though, that these encounters can extend beyond recreational and social gatherings. For example, companies that are starting to think about a new workplace future are also thinking about how the Metaverse might affect them.

Many enterprise applications for the Metaverse will likely start as improved consumer tools. Companies can, for example, construct wonderful virtual environments in which to develop goods, test ideas, and collaborate with peers. Virtual reality meeting rooms and interactions augmented by MR and AR are already becoming more prevalent in the workplace. Many firms in many industries are already using extending reality and Metaverse concepts to bring professionals and employees together in a hybrid environment.

People will jump into unique situations where they can establish muscle memory and learn new skills thanks to the Metaverse, which will provide more great options for online learning and training in the workplace. We might even see the construction of virtual landscapes where people can get more easily accessible tools for enhancing their workflow whether they're at work or not.

Microsoft and Accenture have already experimented with creating virtual workspaces for collaboration. At the same time, Meta's Horizon Worlds provides virtual workspaces via headsets.

Metaverse Media and Entertainment

Because gaming is the foundation of the Metaverse, it's only natural that the entertainment and media industries benefit from it. Many shared virtual events like concerts and gatherings are already taking place in environments like Minecraft, Roblox, and other similar spaces, indicating that entertainment in the Metaverse is becoming a common concept.

The Metaverse was a critical tool for many event organizers in keeping people linked in a world when in-person activities were no longer viable. For example, customers could watch NFL games from the sidelines in virtual reality or visit virtual exhibitions at the National Museum of American History. In Fortnight, we've seen concerts and even world premieres.

Virtual events in the Metaverse will no longer be limited to the consumer landscape. Several professional groups are likely to use the technology as well. In addition, future Metaverse media encounters will merge in-person interactions with digital-twin events in the virtual world to provide more hybrid experiences for people who cannot attend conferences.

Metaverse events in the future will allow us to completely participate in event experiences from afar, communicating with other attendees, purchasing products in real-time, and even sharing information with people we want to connect with in the future. Therefore, the development of capabilities such as 5G will be critical in assuring real-time interactions between in-person and online event attendees.

Metaverse Economy

The establishment of new and improved economies is one of the most important opportunities for Metaverse's future. We're already seeing indications of these new marketplaces, where traditional and digital-native firms interact with customers through virtual and augmented reality and the distribution of new digital goods.

Fashion retailers have made a beeline for the Metaverse in a few years. For example, Burberry teamed with Elle Digital in Japan to create a digital version of one of their stores where shoppers could browse and shop. In addition, DressX revolutionized the fashion industry by allowing buyers to try on and buy digital clothing for their online avatars. Companies have even created "digital twins" of things for use in the online realm. For example, in Roblox, gamers can redeem virtual versions of real-life things, such as a digital twin of a Nerf rifle.

The use of real money to purchase digital products and services is perhaps the most basic example of a Metaverse economy. For example, there are many gaming communities where users may buy "loot crates," avatar skins, and virtual goods. One of the most well-known examples of this is Fortnite. Some people even sell their knowledge in a game for an hour or two or give their talents to create virtual art.

The connection between digital and physical economies is becoming increasingly important. For example, people can shop for material things in a virtual world via eCommerce experiences (like the example with Burberry). Real estate companies can use virtual reality to show people around a house before buying it. Travel agents can provide shoppers with an insight into a vacation experience.

The digital real-estate market is commonly seen as

promising to build the Metaverse economy. Customers are already participating in the Metaverse's real-estate purchase and sale. People can even buy and sell houses to make money in an AR game from 2020. In addition, early brand involvement in collaborations and sponsorships demonstrates the importance of this Metaverse economy. For example, customers may purchase copies of the North Face's clothes in Pokémon Go, and Wendy's has its branding in Fortnite.

The gaming industry sparked the match to ignite the growth within the Metaverse. However, the social, business and commercial industries will continue to expand on their infrastructures, and I do not doubt that the Metaverse will continue to take us by storm.

THREE TECHNOLOGY PILLARS WITHIN THE METAVERSE

Today, the Metaverse's promise is centered on universality and decentralization. It invites us to envision a world where we have greater control over our digital experiences and may access them more flexibly. As the "new internet," the Metaverse is still in its early phases of development, with new technology constantly improving the environment. Blockchain technology, AI, and Extended Reality (XR) are the three most common forms of technology linked to the developing Metaverse. Here is a little more information about them.

Blockchain technology

The blockchain is a vital component of decentralization. It is no longer only a concept linked with Bitcoin and cryptocurrency. In the Metaverse, blockchain promises to give individuals greater control over their online experience, ushering us away from the static Web 2.0 and into Web 3.0, where larger corporations such as Google and Amazon have less control over what we do and see. With technology like NFTs to invest in and support artists, smart contracts, and decentralized finance, blockchain technology is already making waves in the Metaverse.

Artificial Intelligence (AI)
Improving the relationship between the physical

and digital worlds necessitates a certain level of machine intelligence. A variety of Metaverse experiences require Artificial Intelligence. In addition, it can aid in natural language processing, ensuring that our machines and robotics understand us. AI also supports computer vision and Simultaneous Location and Mapping technologies, which enable robots to grasp our physical environment.

Extended Reality (XR)

XR is perhaps the most commonly referenced form of Metaverse technology today, involving blending the physical and digital worlds through headsets and devices. We can enter virtual worlds and interact with 3D avatars in communities using extended reality. Mixed and augmented reality can also bring digital content into the real world, transforming how we engage with everything from maps to shopping experiences.

THE RISE OF THE METAVERSE DIGITAL ECONOMY

There is plenty of potential for the digital economies of the Metaverse to evolve as our virtual and physical worlds become more aggressively interwoven. For example, as more people join groups using virtual or digital avatars of themselves, there's a significant probability that more "virtual items" will acquire traction.

When you buy a high-end designer gown, it may come with a digital twin that you can wear with your virtual avatar. In addition, purchases made in the digital realm can impact our physical belongings differently. For example, a person might buy a handbag at a virtual store and have it delivered to both their avatar and their real-world home at the same time. While there are many ways to see the Metaverse economy evolve, the creation of non-fungible tokens (NFTs) is likely a fascinating example right now.

The potential for everyone to have more influence over their digital landscape through a decentralized and open environment is a key component of the Metaverse. Instead of allowing our ideas to fall into the hands of larger businesses, creative workers are pushing for more ways to commercialize them in this context. NFTs are a critical component of this fight.

NFTs are a concept from the Metaverse's "blockchain" pillar. Customers can use an NFT to verify their digital items ownership and support digital artists and innovators in new ways. Unfortunately, due to the structure of Web 2.0 (our current online ecosystem), with the majority of any money and advantages being split among a few significant entities, such as Google or Amazon, artists have often struggled to market their efforts. Creators will have greater opportunities in the Metaverse to own and share their important tools without a middleman getting a cut.

The Metaverse relies heavily on user intention and agency, and individual people's innovation has already boosted engagement in various settings. Roblox, for example, is based on the capacity for regular people to create and distribute games. Over 9.5 million developers on the platform, and user-generated content accounts for more than half of all in-game currency spent worldwide. This track record provides insights into how the Metaverse and Web 3.0's developing economy might provide ordinary people with more power and earning opportunities.

The forces behind Metaverse's rise

Because technological innovation is occurring at a higher rate than ever before, the Metaverse is accelerating at an astounding rate. In addition, we've been obliged to invest substantially in our digital lives to supplement and augment limited physical environments, thanks largely to the pandemic. The development of different technology lynchpins will make the Metaverse more accessible to everyone as we move further into the Metaverse.

Virtual, augmented, and mixed reality

Users can enter the Metaverse through virtual reality (VR), bridging the perceived divide between digital and physical realms. We can explore new places and make experiences more accessible to everyone by using virtual copies of people, objects, and landscapes. VR allows people to walk into events, shop at stores, and learn about new things.

But on the other hand, augmented reality (AR) and mixed reality (MR) will allow us to improve our real-world experience like never before. Even haptic feedback tools, which are part of the extended reality (XR) landscape, will allow us to bridge the gaps in our relationships and feel our contacts' handshakes and hugs wherever they are.

Improved interconnection

Minimizing the gap between our digital and physical worlds is essential for creating an immersive Metaverse experience. In the future, solutions like 5G, which allow for more immersive virtual and augmented reality experiences, will be critical. In addition, the Metaverse will require better connectivity and more efficient use of digital capacity.

We'll also need to figure out how to manage the massive volumes of data that must be processed every day in this new environment in an environmentally friendly and efficient manner. Companies are already researching innovative ways to generate sustainable energy to power the Metaverse and its experiences.

Landscapes development

The Metaverse is an open, interoperable, universal ecosystem designed with humans. This will necessitate quick access to development prospects for all creatives and builders. In addition, Low-code and no-code technology choices allow more people to access future Metaverse prospects.

Furthermore, businesses are increasingly deploying open and accessible systems to produce extended reality landscapes. In a perfect world, these open environments would be standardized and interoperable, allowing any section of the Metaverse to connect with any other without fear of functionality conflicting. But unfortunately, experts expect that full standardization and interoperability will take some time.

CREATING AN ETHICAL METAVERSE

An ethical and sustainable Metaverse will need to be developed on transparent foundations. The idea of complete transparency and decentralization is already a big part of what makes the Metaverse and the Web 3.0 philosophy beneath it so intriguing. The Metaverse will be built on decentralized and distributed ledgers, allowing transactions to be verified and increasing our trust between one another. Therefore, the future Metaverse will have to be inclusive, safe, and innovative.

Inclusive

One of the Metaverse's distinguishing characteristics is that it should always be human-centered and accessible to everyone. The technologies that emerge in this new environment must be approachable and simple to comprehend. Everyone should have the opportunity to take part in this new world.

For example, suppose a person using a wheelchair in the physical world cannot enter a building because the ramp is too steep. In this case, the building is inaccessible to all. In another example, suppose a person with motion sickness uses a VR headset to enter the Metaverse. In this case, the person might get sick after 30 minutes in the Metaverse. Thus, the developing headset technology is compromising accessibility for everyone.

Leaders and designers of the Metaverse will need to be inclusive by design, the leading form of human-centered design. This design style focuses on minority audiences often overlooked in the design process. It aims to improve innovation outcomes and value for all customers.

Safe

It's no surprise that the Metaverse is gaining traction as a secure, private, and secure environment for the digital future. Because of blockchain and decentralized concepts, fraud and other concerns are much more difficult. However, as inhabitants of the future Metaverse, it is still our responsibility to make the environment secure and safe.

Unfortunately, bad actors will always find a way to exploit systems. Just as previous leaders dealt with fake accounts, bullying, and harassment on social media platforms, leaders will face several challenges in the Metaverse. Examples include identity theft, invasion of personal space, bullying, harassment, abuse, and many others.

Innovative

The Metaverse, like any technology, should have plenty of room to grow and expand as we learn more about it as a species. Already, we're seeing a slew of new Metaverse potential emerge from the creation of new XR tools and NFTs. A powerful Metaverse in the future will constantly be learning and growing.

It's not just about producing the most immersive VR experiences or making the web more decentralized when it comes to building a strong Metaverse for the future of technology. Instead, taking three steps forward and two steps back enable humans to employ technology and advance ethically and sustainably.

WHAT DOES THE METAVERSE MEAN FOR US?

Hopefully, this look at the Metaverse today and where it's going has given you a clearer idea of what this environment might imply for humans and our future. The Metaverse is a virtual world where we can learn, create, play, communicate, and interact with everyone. It makes the world smaller and links us together regardless of our geographical location. It might be the start of new contexts to create entirely new economies based on value sharing. Fans of the Metaverse see it as the start of a new era. People create and enjoy experiences in an increasingly accessible landscape.

However, there are possible difficulties with the Metaverse, such as the risk that we may struggle to trace down and prevent crime in a decentralized environment. Therefore, it is up to us as a species and a global community to harness the Metaverse's potential for good. Finally, the environment can be beneficial if we focus on creating a Metaverse for everyone and accessible to everyone. This new technological advancement is merely another instrument towards a better future; how we use it is up to us.

HOW TO INVEST IN THE METAVERSE

Now that you know more about the growing Metaverse industry, you might consider investing in it. Of course, you'll need a web wallet for many of these investment strategies, for example, Metamask. You will also need to know how to interact with Web 3.0 applications. Because the landscape is changing so rapidly, I will not be providing step-by-step instructions. Instead, it would help if you did your research from the official company's websites and social media channels.

Please remember that this is not investment advice, and every decision is yours to make. Seek advice from a financial advisor and always do your research. Here are the best ways to invest in the Metaverse.

Digital cryptocurrencies (best exposure, good returns, volatile)

One of the best ways to get exposure to the Meta-verse is to purchase digital cryptocurrencies. Each crypto project will use them differently, but most use them for in-game purchases, governance, and project ownership (similar to holding a company's share). Some examples of the most popular digital cryptocurrencies are Axie Infinity ($AXS), Decentraland ($MANA), The Sandbox ($SAND), Enjin Coin ($ENJ), and Radio Caca ($RACA). To obtain these digital cryptocurrencies, you can purchase them directly from centralized exchanges such as FTX and Binance or via decentralized exchanges such as Uniswap or Sushiswap.

Suppose you're not sure which project is going to

outperform others. In that case, you can invest in a Metaverse index akin to an Exchange Traded Fund (EFT). For example, Metaverse Index ($MVI) by Index Coop is a collection of 14 underlying tokens designed to capture the trend of entertainment, sports, and business within the Metaverse. By investing in a basket of assets, you can reduce the volatility with individual projects and apply your investment thesis more broadly across the Metaverse sector.

Alternatively, suppose you want to invest in a more fundamental and underlying asset. In that case, you can purchase the layer one smart contract projects on which the Metaverse is built upon. For example, Decentraland ($MANA) is built on Ethereum ($ETH). You could get indirect investment into Decentraland by purchasing some Ether.

NFTs and in-game items (high risk, high reward, illiquid)

Another way to get exposure to the ever-growing Metaverse industry is to purchase NFTs and in-game items. Again, each crypto project will use them differently. However, investors use them for in-game interactions, showing off rare collections, access tokens to private events, or cosmetic purposes.

You'll first need to figure out what project you want to invest in and where to buy the digital asset. Next, you will need to get a web wallet, for example, Metamask. Then, you'll need gas for the blockchain you are on, for example, $ETH for the Ethereum blockchain or $BNB for the Binance Smart Chain. Finally, you can find your digital asset of choice on the project's website or a secondary marketplace—it depends on the project and the release time.

Virtual land (high risk, high reward, illiquid)

Another fascinating way to get exposure to the Metaverse is to purchase land and digital real estate in the form of NFTs. We can buy, sell, and even rent land in the Metaverse like physical-world real estate investors. But, again, you'll need a web wallet, cryptocurrency for gas on the blockchain, and cryptocurrency to exchange for the NFT. You can purchase real estate in The Sandbox, Decentraland, Bloktopia, and other upcoming projects. However, just like the physical world, the land is finite and scarce, so entry-level prices are generally quite high—the floor price (minimum price) of land is currently sitting at around 3 Ether or $10,000.

Metaverse Stocks (safest bet, low returns)

Metaverse-associated equities are stocks that are actively involved in Metaverse development. For example, the companies could be involved in creating virtual reality (VR) goggles, networking technology, 3D rendering apps, and more. Investors can purchase these stocks through brokerages or ETFs that track the Metaverse (Exchange Traded Funds).

Investing in publicly traded companies whose business strategies or profitability are related to the Metaverse is the least volatile choice for retail investors looking to purchase into the Metaverse. The following companies are some examples:

Meta Platforms Inc (NASDAQ: FB)

This company develops and sells social media platforms. CEO Zuckerberg recently stated that the firm, formerly known as Facebook Inc., would be rebranded as Meta Platforms Inc. Since the announcement, Meta has released Horizon Worlds, a virtual reality Metaverse platform. In addition, the Oculus Quest 2 VR headset from Meta was also one of the most popular recent holiday gifts.

Roblox (NYSE: RBLX)

Roblox is an online Metaverse platform that allows users to create and share virtual environments. Since its debut in 2006, Roblox has grown significantly, with 9.5 million developers, 24 million digital experiences, and 49.4 million daily active users—an increase of 35% year over year. Despite these figures, the business has yet to make a profit.

Boeing (NYSE: BA)

This company manufactures airplanes. Boeing is utilizing the Metaverse to enhance and expand its manufacturing capabilities. In an interview, Boeing's top engineer said that the jet maker plans to develop a proprietary digital environment where its human, computer, and robot employees can communicate and collaborate seamlessly.

Microsoft (NASDAQ: MSFT)

This is a software company based in Redmond, Washington. Microsoft is attempting to carve out a Metaverse niche in the business sphere. Individuals will create personalized avatars and work in a holographic

3D environment crossing geographic boundaries. Holo-portation, a mechanism that allows users to access the previously described digital environment with a VR headset, will be a fundamental element of Microsoft Mesh. The user appears as a lifelike digital representation of themselves, with the ability to interact with team members in real-time.

I hope you enjoyed this section on how to invest in the Metaverse. I am adopting a diversification and dollar-cost averaging strategy for fundamental crypto projects like Bitcoin and Ethereum. I ride the waves of the Metaverse up and take profits early. The dollar's value is decreasing with the rise of inflation, and crypto is the number one ultimate way to hedge against (and indeed beat) this never-ending rat race.

AFTERWORD

Although the Metaverse is still a long way from being a reality, the pieces are starting to fall into place. Meta and other internet titans might experience initial success with their earliest prototypes of virtual places. However, with the growth of blockchain technology, machine learning, and artificial intelligence, the real success will come from companies that integrate all aspects of the future Metaverse.

The future Metaverse will be a vast network of interconnected, real-time rendered 3D environments with a sense of continuity in terms of digital identity, objects, transactions, and payments, among other things. It will also be possible for many users to experience it seamlessly and simultaneously.

Companies building towards the future Metaverse are spending millions of dollars to persuade customers that the Metaverse has arrived. Will it, however, bring in a new era of broad acceptance and barrier-free digital contact? Or will it remain a niche product for gamers and future tech enthusiasts?

The myriad potential linked with blockchain technology will profoundly impact the future Metaverse. The Metaverse provides a digital environment where anything is possible and non-fungible token (NFT) technology provides verifiable ownership and uniqueness. Therefore, NFTs will forever change our social and economic experiences if the physical and digital worlds combine.

The Metaverse began with gaming, and many enthusiasts want to get on board before it's too late. NFTs enable players to take control of their gameplay experiences. And although many of the Metaverse games are still in their infant stages, their intriguing gameplay experiences and unique features attract countless new players every day.

Play-to-earn NFT games have also swept the crypto community from their feet. Games such as Axie Infinity have attracted nearly 3 million monthly active users, growing rapidly. In addition, bounties such as NFTs and cryptocurrencies are driving players insane. Players can now monetize their time and achieve real-world value by exchanging bounties for local fiat currencies.

The Metaverse promises to be an exciting new chapter in the internet's history. Our physical and virtual lives will merge more than ever before—virtual land will be a vital component of these new virtual worlds. Twenty years ago, a Fifth Avenue storefront was the holy grail. Now, a first-page Google or Amazon ranking may bring millions of dollars. And tomorrow, a premium piece of virtual real estate in the appropriate virtual environment could be the equivalent. However, because this is a new field, there are many dangers, and we're still early in the adoption curve. As a result, it's still a gamble about which platforms will emerge as the final victors and what people desire.

It is widely assumed that you will not need to use a virtual reality (VR) headset to enter the Metaverse. However, you will have access to the entire profound experience if you do. In the future, business people may utilize VR headsets for jobs currently accomplished with cell phones. The Metaverse, on the other hand, will not be limited to VR and will be available via augmented reality (AR) devices as well as other Internet-connected gadgets. Therefore, AR will be a massive component of the future Metaverse. However, more technological advancements are required to push this narrative.

The Metaverse is an extension of the static Web 2.0 internet in terms of scale. There will be a demand for permissionless identification and financial services behind the scenes of the Metaverse. In addition, billions of individuals will need to store data in the Metaverse. Blockchain technology has the key to solving these complex issues.

Blockchain companies such as Decentraland and The Sandbox have already created virtual worlds that integrate cryptocurrencies and digital assets. They have enabled gamers to design and monetize virtual experiences and services. Likewise, humans will use virtual spaces to engage, explore, communicate, do business, and even establish communities in the blockchain Metaverse.

Finally, the blockchain Metaverse is not merely a science fiction concept. If you look around closely, you will realize that everything is becoming digitized. While it is critical to comprehend the Metaverse's value benefits, we must not overlook the problems. The Metaverse has been a long time in development, and new technologies like blockchain have accelerated this process. Welcome to the Metaverse.

RESOURCES FOR YOU

Here is a list of the most important resources you can find in the crypto world. For full transparency, please understand that some of these links are affiliated. Even if you don't want to use the referral links, check out their websites because they truly are the best in the business.

BUY AND SELL CRYPTO ON CENTRALIZED EXCHANGES

FTX - 5% off trading fees - Buy and sell BTC, ETH, SOL, and index futures (+ many more) with low fees and up to 20x leverage - https://ftx.com/referrals#a=BMETA

Binance - 5% off trading fees - Buy, trade, and hold 600+ cryptocurrencies - https://accounts.binance.com/ en/register?ref=MY2U7UEA

STORE YOUR CRYPTO SECURELY

Ledger Nano X - The secure gateway to your crypto needs - https://shop.ledger.com?r=cc2387e9faa8

EARN INTEREST ON YOUR CRYPTO

Celsius - Sign up and earn $50 in Bitcoin - Buy, borrow, swap & earn crypto with some of the best lending interest rates - https://celsiusnetwork.app.link/144620c21f

MICRO INVESTING

Bamboo - Sign up and earn $10 - Start investing a little every day - https://app.getbamboo.io/invest and enter code 800c816

ONE LAST THING BEFORE YOU GO...

Thank you for purchasing The Blockchain Metaverse. I hope you enjoyed it. If you did, would you consider posting an online review? Your feedback will help me create better quality books, and it will also help others make more informed decisions.

To leave a review, search for The Blockchain Metaverse on Amazon (or on your favorite online retailer of choice) or find this book in your recent purchases.